AF334080

No. 828
$9.95

SWITCHING REGULATORS & POWER SUPPLIES

with Practical Inverters & Converters

By Irving Gottlieb

Preface

The widespread use of regulated power supplies based upon a new principle—the principle of *minimum power dissipation*—is now an accomplished fact. It has long been almost a sacred tenet of engineering philosophy to strive for the highest possible efficiency in any system. Thus, the traditional and relatively inefficient *linear* or *dissipative* power supply has been a sore point for many years. It makes sense to have dissipation in an electric heater—but certainly not in a power supply.

The advent of the switching-type power supply has made it possible to significantly decrease the dissipation in power supplies, along with the accompanying bulk, cost, and thermal problems. Indeed, the theoretical efficiency of a "switcher" is a cool 100%—corresponding to *zero* dissipation. Yet the implementation of the switching principle has had to wait until the technological state of the art was able to create suitable components for these power supplies. In the past, only the military and aerospace industries were able to afford such high-performance power supples. But now the technology has advanced enough and the cost has dropped enough that switching-type power supplies are no longer considered to be something exotic and untouchably expensive.

There are still some fears and doubts concerning switchers today. Some of these fears have legitimate roots—complexities, cost, electrical noise—but the modern switching-type power supply has overcome most of these early obstacles and shortcomings. New semiconductor devices, capacitors, magnetic components, and integrated circuits have made it possible to put a switching-type power supply together in almost "cookbook" fashion. The new components have greatly increased efficiencies, reduced costs, and simplified designs.

Yet the biggest obstacle to the switching-type power supply has been the reservations of engineers, technicians, and designers to try something new. Many doubts have been raised in the past, and these are difficult to suppress or explain away. But the race to produce and utilize these highly efficient power supplies, to increase the efficiency of a system, has ultimately won over such doubts. Now these supplies may be found in such diverse applications as instrumentation, transportation, consumer products, and hobbyist electronics, as well as military and aerospace applications. Today, there is a need for engineers, servicemen, technicians, experimenters, and countless others to know and to understand how these new switching-type power supplies work, so that they can design them and use them to a profitable advantage.

It was my good fortune to have access to the expertise of many pioneers in this intriguing field of electronics. Mr. Robert Boschert, president of Boschert Associates, provided valuable insights, as only an original advocate and manufacturer of switching-type power supplies could provide.

Inasmuch as a central objective was to depict applications of recently developed devices and circuit techniques utilized in making power supplies using switching principles, my goals could not have been achieved without the cooperation of important research-oriented firms in this field. Acknowledgment of the following companies, as well as gratitude to the specifically mentioned individuals are accordingly extended as follows:

Mr. W. C. Caldwell, Supervisor, Customer Services Development—National Semiconductor Corporation.

Mr. Robert Dobkin, Director, Advanced Circuit Development—National Semiconductor Corporation.

Mr. Forest B. Golden, Consulting Application Engineer—General Electric, Semiconductor Department.

Mr. Lothar Stern, Manager, Technical Information Center—Motorola Semiconductor Products, Inc.

The activities and accomplishments of these people have been both competitive and mutual. The products of the companies they represent have been both unique and overlapping in the marketplace. And this cumulative effort and combined experience have done much to make the switching-type power supply a reality today. It is my sincere hope that this book will enhance the appreciation and promote the use of this unique type of power supply.

Irving M. Gottlieb
Menlo Park, California

Contents

1

A Power
Supply Switch

Modern switching-type power supplies are characterized by the use of semiconductor devices that actually switch or interrupt the flow of current within the supply. While this may sound like a rather odd mode of behavior, there are many benefits to be derived from such operation that cannot be obtained by traditional methods. These major benefits include high efficiency, small size, and the inherent capability to operate from a much wider range of input voltages. Additionally, as the cost of power semiconductors and integrated circuits continues to decrease, switching-type power supplies may also gain a very significant economic advantage.

Before embarking upon a detailed investigation of switching-type power supplies, it would be wise to become acquainted with the subject of power supplies in general. What, for example, is a power supply? What are we really dealing with? It is essential that we pin down a few definitions at the very beginning to avoid later confusion. Even the nomenclature is cause for confusion, for a *power supply* does not usually generate or produce the power in itself, but rather obtains the electrical power from a utility company located miles away. Rarely, when referring to a power supply, are we

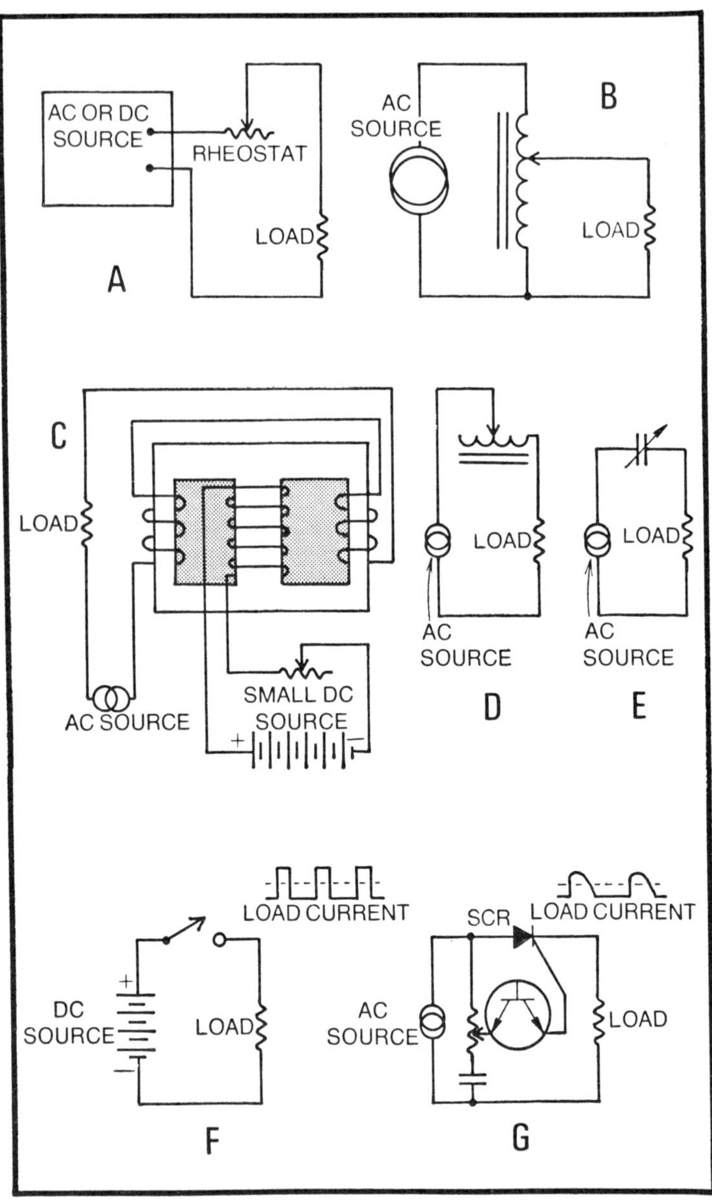

Fig. 1-1. Basic schemes for controlling electrical power supplied to a load. Method A is probably the most common technique, but is inherently the least efficient. All of the other methods are, at least in their ideal form, nondissipative. Although the switching technique depicted in F appears to be the most primitive type of control, it is the basis for modern high-efficiency power supplies.

ever speaking of a battery or generator as the prime source of power—and even when we are, we most often exclude the battery or generator from our thoughts and concentrate solely upon the electrical circuitry that is connected to it.

WHAT IS A POWER SUPPLY?

Electronic power supplies (or power supplies used in electronics) are most often charged with the task of *altering, controlling,* or *regulating* electrical power. The word *control* is quite inclusive: whether we rectify, invert, regulate, or change the voltage or current levels, some control technique must be involved. Such control of power delivered to a load can be achieved only by absorbing surplus power in the control device. In this time of energy consciousness, the very concept of *surplus* or *throwaway* power deserves close scrutiny.

On the other hand, there are techniques which allow the control of load power *without* power dissipation in the controlling device. The basic methods of controlling electrical power to a load are shown in Fig. 1-1. Perhaps others may suggest themselves, but they will most likely be recognized as versions of those depicted. For example, there are other ways of using saturable-core devices, or magnetic amplifiers (usually shortened to *magamps*), and the rheostat is representative of many other devices, such as transistors, tubes, and thermistors.

WHAT IS A SWITCHING-TYPE POWER SUPPLY?

It is difficult to give a quick and concise definition that would distinguish switching-type power supplies from all other types. But it would be appropriate to say that a switching-type power supply is one in which the main flow of electrical power is generated, controlled, or regulated by means of switching devices. Most often, the switching-type power supply has an electronically regulated output. Switching-type supplies are usually more costly, and are thus expected to justify their performance, reliability, and cost on a comparative basis with conventional linear or dissipative-type power supplies.

Some power supplies involve a switching process, but are not ordinarily referred to as switching-type supplies. The

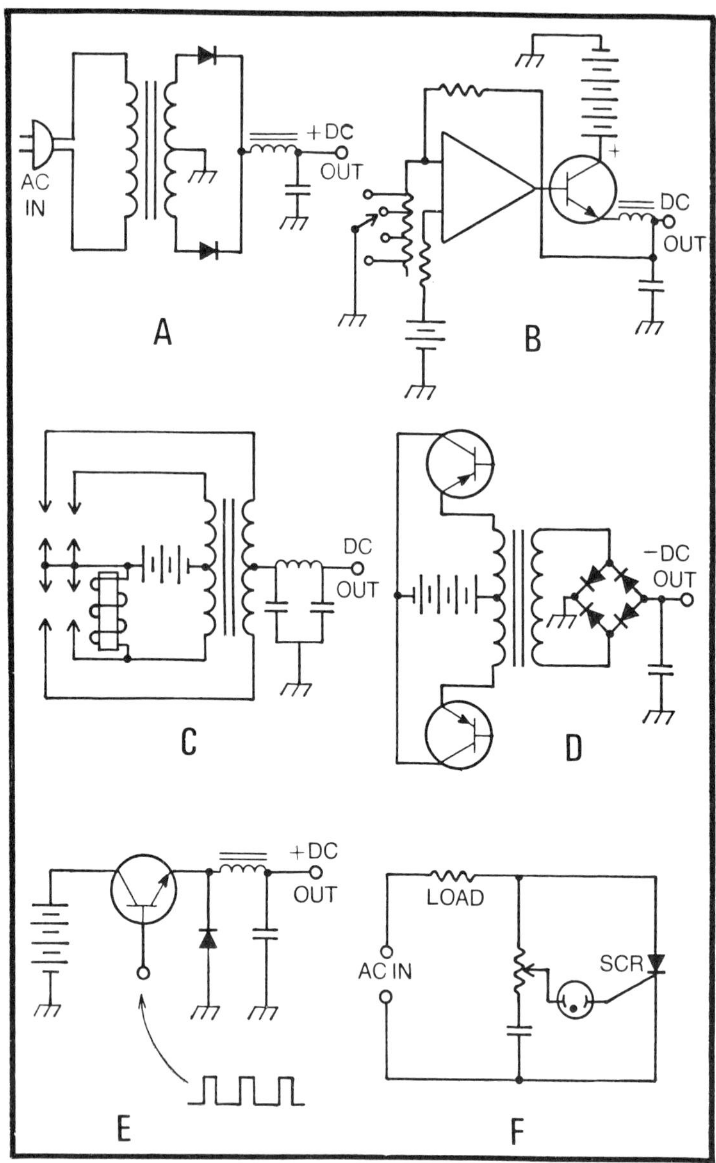

Fig. 1-2. Various supplies in which switching techniques are used. From an academic viewpoint, the ordinary rectifier circuit of A must be classified as a switching power supply. In practice, however, rectifiers do not qualify as a switching-type power supply. The once-popular vibrator supply of C and its modern version, the saturable-core converter of D, qualify as switching-type power supplies. The switching transistor supply of E, and the thyristor supply of F are also switching-type power supplies.

reason for this amounts to nothing more than custom or tradition. A good example is the rectifying circuit shown in Fig. 1-2A, which from an academic standpoint would technically qualify as a switching-type power supply. How else can rectification be accomplished except through switching?

On the other hand, the mere fact that a power supply incorporates a switching process does not necessarily qualify it as a switching-type supply. For example, there are many new *programmable* power supplies of the linear type that utilize switching elements to establish the output voltage of the supply by switching resistances to alter the internal reference voltage, as illustrated in Fig. 1-2B. Other supplies, such as the old-fashioned vibrator type, generated and often rectified voltage using elements that are most definitely identifiable as switches.

The last three power supplies illustrated in Fig. 1-2 are more readily accepted as switching-type supplies by modern-day standards. In the popular converter, shown in Fig. 1-2D, the switching process is clearly evident. Switching in such supplies is usually performed by transistors or thyristors, often in conjunction with a saturable-core transformer that performs the principal switching task. Switching regulators often take the form illustrated in Fig. 1-2E, in which the transistor, diode, and inductor perform the major switching and regulating functions. For AC voltage control, regulation is accomplished by a thyristor switch, as in Fig. 1-2F.

Figure 1-3 illustrates a switching-type regulator in somewhat more detail. Such regulators are powered by an unregulated DC source. The switching process is used to perform the regulatory function by interrupting the flow of current from the unregulated supply. The operation of the switching device is controlled by an error amplifier or comparator, a feedback loop in the system which continuously compares the output voltage to a reference voltage and automatically adjusts the switching operation to obtain the desired output voltage.

WHY IS A SWITCH BETTER THAN A RHEOSTAT?

In *dissipative* regulation of voltage or current, power is deliberately thrown away. The dissipative element, often a

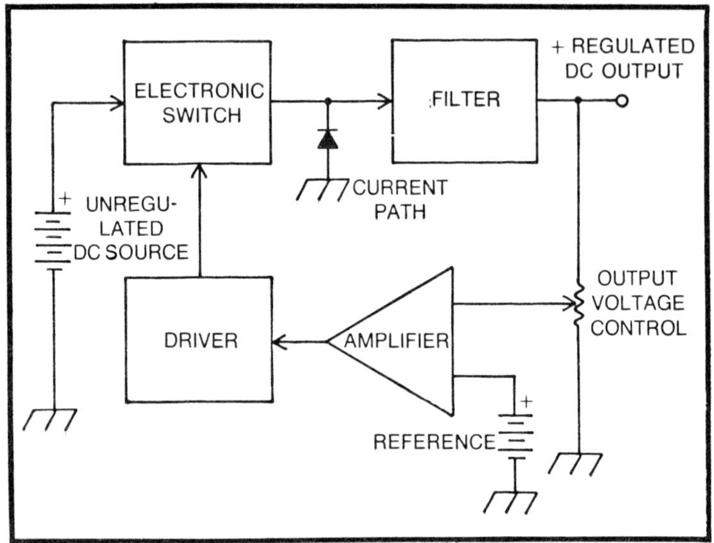

Fig. 1-3. Most generally, a switching-type power supply is a voltage regulator. In the block diagram above, the electronic switch may be a transistor or thyristor, or a pair of these devices in an inverter circuit. The filter, in conjunction with the current-path diode, may be more descriptively designated as an energy-storage system. The amplifier is in the feedback path and may also perform the function of a voltage comparator. The driver may be a monostable multivibrator, a Schmitt trigger, or an emitter follower.

power transistor, is imposed with the task of "soaking up" the excess power, and this naturally gives rise to heat-removal problems. It does not matter how sophisticated the control electronics are in such a power supply, the fact still remains that the dissipative element functions as a rheostat—a relatively crude method of regulating power.

In the switching-type power supply, a switching device is substituted for the dissipative device. Control or regulation of power is then achieved by varying the *duty cycle* or *repetition rate* of the switch, rather than its *resistance*. It is certainly valid to wonder what is so wonderful about this technique, considering that it must obviously be more complicated. The answer is that an *ideal* switch does not absorb or dissipate power—it is completely *on* or completely *off*, with no intermediate resistive state to dissipate power. As a result, the overall efficiency of a switching-type power supply is usually higher—much higher—than conventional dissipative-type power supplies. And this factor is becoming an overwhelming point in favor of switching-type supplies, because it results in significant advantages in size, weight, energy conservation, and cool operation. When these advantages are examined in relation to the requirements of the overall system, they often produce a significant economic advantage as well, particularly for high power levels.

There are a number of advantages that may be attributed to switching-type power supplies. But to be truly meaningful, the switcher must always be compared to a nonswitching, linear supply having similar power ratings. When this comparison is made, it will usually be found that the switcher has certain natural advantages over dissipative types, which may be summarized as follows:

- Higher electrical efficiency.
- Lower operating temperature and relaxed heat-removal problems.
- More compact packaging.
- Lighter weight.
- Wider range of input voltage.
- Better "coasting" ability for momentary interruptions of the AC power line.

These are the basic superiorities of the switching process. There are other parameters and characteristics such as cost, reliability, and output purity, which will be discussed later. And, as may be suspected, there are also a few shortcomings. As with other circuits and systems, the switcher displays beautiful performance in one domain at the expense of poor behavior in others. There should be no delusion that the switching supply presents a panacea for all electronic ills. It does happen to be true, however, that its virtues often override its flaws, and this fact has resulted in the displacement of the dissipative supply by the switching type in an ever increasing number of applications. The switching supply has been a theoretical possibility for a long time, but only recently has it blossomed into practical hardware. This has been brought about by a combination of factors: semiconductor evolution, improved components, and advanced circuit techniques. The

**Table 1-1. Generalized Comparison Between
Switched and Linear DC Regulators.**

FEATURE	SWITCH TYPES	LINEAR TYPES
Efficiency	65% to 85% is common.	25% to 50% is common.
Temperature rise	20 C to 40 C is readily achieved.	50 C to 100 C is not uncommon; depends greatly upon heat-removal techniques.
Ripple	20 to 50 mV peak-to-peak is often encountered. Smaller ripple voltage is usually difficult to achieve.	5 mV peak-to-peak is not difficult to attain and lower values can be had at greater cost.
Overall regulation	0.3% is common specification. Tighter regulation is usually difficult to achieve.	0.1% is commonplace and much tighter regulation is available at greater cost.
Weight	30 watts per pound is common.	10 to 15 watts per pound is often encountered.
Volume	1 cubic inch per watt tend to be rule of thumb for many designs.	2 to 3 cubic inches per pound depending greatly on heat removal method.
Isolation from line transients	Very good. often greater than 60 dB.	Generally inferior to switching types. Noisy line often plagues load.
RFI and EMI	Can be troubleshome. Requires attention to shielding. suppression and filtering.	Less likely to be adverse factor.
Magnetics	Some designs can dispense altogether with bulky 60 Hz magnetics.	Bulky and expensive 60 Hz magnetics is evident in larger ratings.
Reliability	More parts. but recent designs capitalize on enhanced reliability obtained from cooler operation.	Higher operating temperature often degrades reliability.
Cost	Although still generally greater than linear type. the difference narrows in larger ratings. There is general tendency for decreasing cost as new semiconductors evolve.	Small linear types enjoy a cost advantage. However. with all the factors in the overall system considered, other cost factors become very significant in larger ratings.

data presented in Table 1-1 is not all-inclusive, but it does provide initial guidance for the comparison of linear and switching-type supplies.

THE PRINCIPLE OF ENERGY TRANSFORMATION

The goal of a near-dissipationless power supply is easily attainable if the requirement for regulation is removed, for then a simple transformer would be adequate for AC voltages, and an inverter or converter would be adequate for DC voltages. A transformer actually consumes very little power when properly designed, and its output voltage can be easily established by merely varying the ratio of turns between its primary and secondary windings. Similarly, an inverter or converter utilizes a high-frequency transformer whose turns ratio can also be varied.

The requirement for *regulation* of the output voltage or current, despite variations of the input voltage, greatly complicates matters. Regulation normally involves an error-correction process—circuits that can monitor the output voltage and make the necessary corrections to keep that output voltage within specified limits. A transformer does not lend itself to this type of operation, since it is difficult to vary the turns ratio of the windings; there are mechanical methods of doing this, as in a Variac or variable autotransformer, but fast, precise regulation demands a purely electronic process. In effect, what is needed is an all-electronic process for converting energy at one voltage to energy at another, and ideally this process should accomplish this energy transformation without introducing an appreciable amount of energy dissipation.

The modern switching regulator fulfills this demand through a unique switching process that provides for the temporary storage of energy in an inductive element, like that shown in Fig. 1-2E. In operation, the inductor is made to store energy during the time that the series switch is *on*, and to release that stored energy during the time that the switch is *off*. The output voltage or current of the switching regulator is then made controllable by simply varying the duty cycle or frequency of the switching process—a relatively easy task to implement electronically.

This energy-transformation principle may be illustrated by first examining the operation of the simple half-wave rectifier circuit in Fig. 1-4. The rectifier diode actually performs a switching operation in that its function is identical to that of a switch synchronized to turn *on* when the energy to the load consists of voltage pulses having positive polarity.

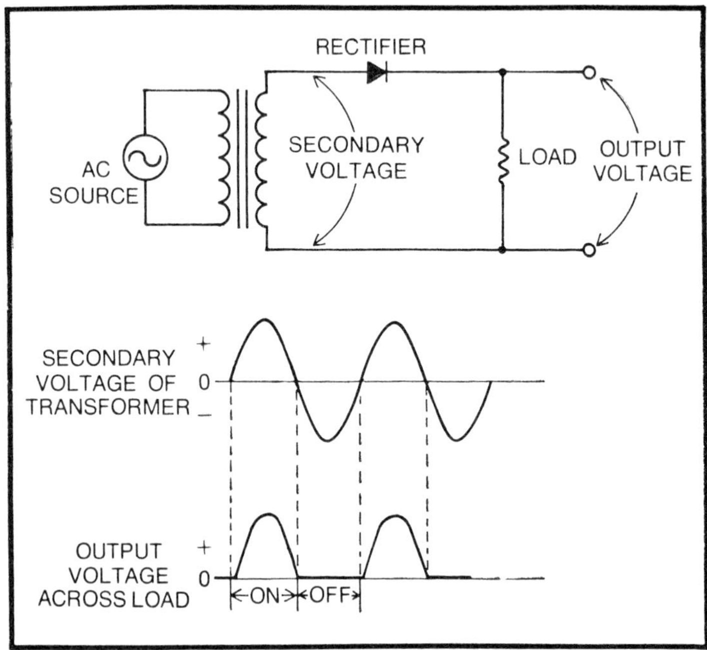

Fig. 1-4. A simple switching circuit. The switching action of the half-wave rectifier is synchronized to the zero-crossings of the secondary voltage. The diode is assumed to be ideal with zero contact potential, zero forward resistance, and infinite reverse resistence.

If an inductor is now inserted in series with the load, as in Fig. 1-5, the energy pulses delivered to the load are stretched out, so that they persist for more than a half-cycle of the applied AC waveform. With the switch in the circuit open, the pulses can never overlap to form a continuous voltage to the load. But with the switch closed, the energy stored in the magnetic field of the inductor is released into the load during the interval that the diode does not conduct. By appropriate selection of inductance and resistance values, the energy delivered to the load can be made continuous, with very little ripple.

The size or value of inductance is inversely dependent upon the operating frequency, or to be more precise, the pulse repetition rate. When driven by the 60 Hz power line, a half-wave rectifier circuit produces 60 pulses per second, whereas a full-wave or bridge circuit produces 120 pulses per

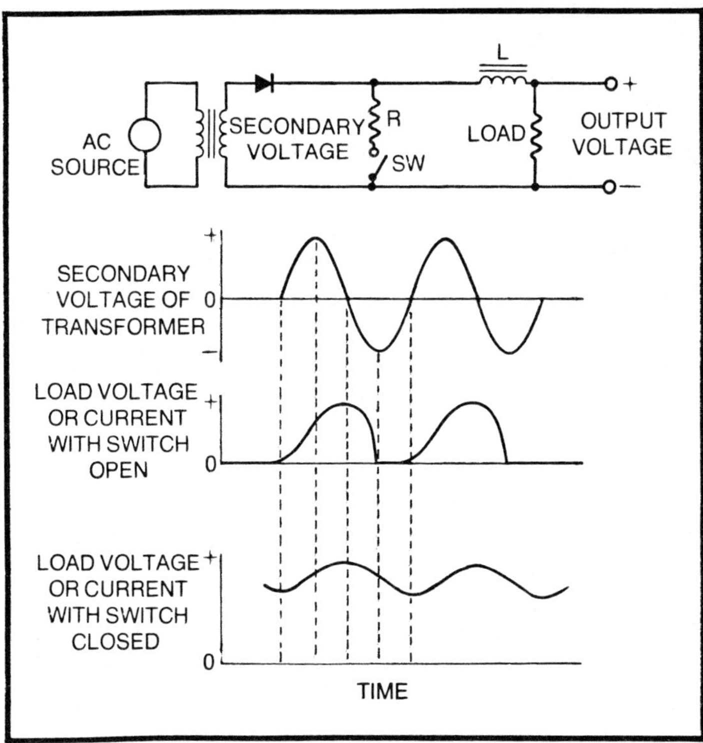

Fig. 1-5. When switch sw is closed, continuous, uninterrupted current flow occurs. This could not be brought about with switch sw open, even if the inductance of L were infinite. This important phenomenon is used to advantage in many switching-type power supplies.

second, or two times the input frequency. (This is a major reason for the preference of full-wave and bridge circuits over the simple half-wave circuit.) The decrease in inductance value becomes quite significant as the pulse rate is raised from 60 Hz to say 60 kHz, for then the inductance value is also accompanied by a reduction in size and cost, though not by as large a factor. But this still serves to illustrate a significant advantage of switching-type power supplies that operate at high switching frequencies.

Returning to Fig. 1-5, the problem with using a simple resistor is that it dissipates energy and so introduces a form of energy loss that defeats our goal of a near-dissipationless power supply. For while the presence of the resistance may be utilized to prolong the release of energy stored in the inductor, it does so only by dissipating some of that energy. Fortunately, this problem is easily overcome.

If the resistor is replaced by another rectifier, the circuit in Fig. 1-6 results. Assuming ideal diodes, the switching operation is entirely dissipationless, since when a diode is conducting, its forward resistance is virtually zero, and when it is not conducting, its reverse resistance is high enough to be considered infinite. Rectifier D_1 imparts energy to the inductor when it conducts, and it only conducts during the time that the AC input voltage exceeds the load voltage. Rectifier D_2 provides a current path for the inductor during the interval that D_1 is not conducting, as is shown by the waveforms in Fig. 1-6. Although the indicated waveforms show that the switching of the current through the rectifiers occurs instantaneously, this would not be so in a circuit using real diodes, for they would exhibit a forward resistance when conducting, as well as a forward voltage drop, and this would tend to slow the switching transition.

The circuit in Fig. 1-6 is representative of both AC and DC switching regulators. If rectifier D_1 is replaced by a controllable switching element, and a DC input voltage is provided, the switching regulator in Fig. 1-2E results. For AC switching applications, rectifier D_1 could be replaced by a thyristor, and output-voltage control could be obtained by varying the phase angle of conduction, in a manner similar to that shown in Fig. 1-2F. In either case, the output voltage is made *controllable* by varying the conduction time of the switching element. Therefore, it becomes possible to regulate the output voltage or current through a near-dissipationless switching process.

REGULATING THE SWITCHING-TYPE SUPPLY

There are a number of circuit configurations, operating modes, and switching devices used in switching-type supplies. It is surprising how many different approaches can be used to attain similar objectives. All of the diverse methods rely upon the all-*on* and all-*off* conductive states in whatever switching device is employed. That is, all switchers avoid partially conductive states. Indeed, this is a good definition of how they

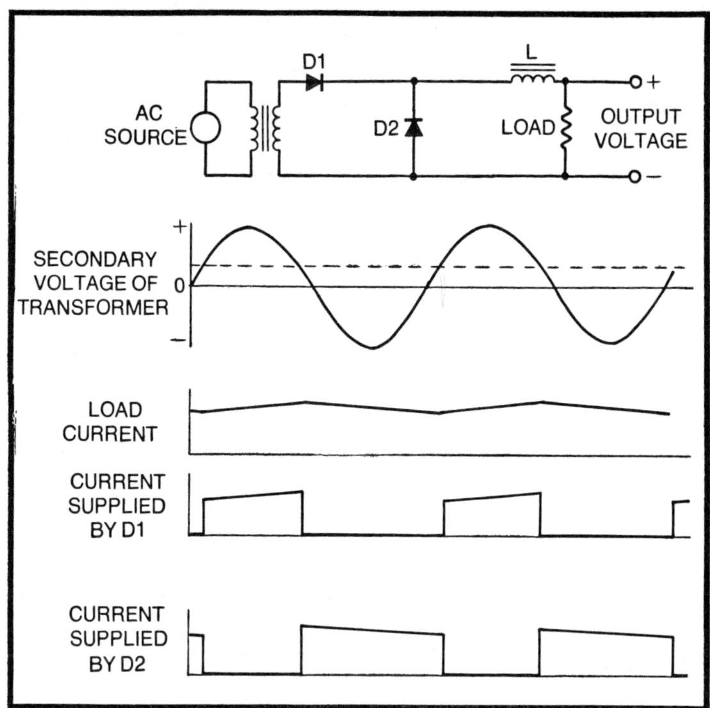

Fig. 1-6. A dissipationless supply results from the elimination of resistive elements. This circuit provides an output voltage that is only about one-third of the peak input voltage, and so would not be practical in conventional power supplies; however, it does serve to illustrate the basic principles utilized in switching-type regulators.

differ from linear dissipative regulating supplies. Even with the singular stipulation that only the two extreme states of conduction are allowed, there is considerable variation in the way that switching can be implemented. Some of these are listed as follows:

- Variation of both frequency and *on* time, either randomly or with constant *off* time.
- Constant frequency with variable *on* time.
- Constant *on* time (or pulse width) with variable frequency.
- Half-wave duty-cycle modulation of a sine wave (as with SCRs).
- Full-wave duty-cycle modulation of a sine wave (as with triacs).

Moreover, the switching process can be implemented with the switching device inserted either in *series* with the load or essentially in *shunt* with it. And the switching can be done at the frequency of the power line, or thousands or even millions of times per second. As might be anticipated, each approach offers unique performance features, tradeoffs, and economic variables. On the other hand, the state of this art is very fluid—the various design techniques compete with one another. The ultimate desirability of one or another approach is often dependent upon the particular development state of certain components. For example, if today one technique appears excessively costly or unreliable because of an inordinately high number of active devices needed, it is quite likely that an integrated circuit will make its debut tomorrow for the express intent of reducing device count to an inconsequential few. The art and technology of switching-type supplies is dynamic and fast moving, so one of the first things we should do is to familiarize ourselves with the various approaches found in these interesting power supplies.

2

Simple Switching Regulators

One of the most important functions of switching-type power supplies is the control or regulation of the supply voltage used to power electronic circuitry. Such supplies are normally required to maintain a fairly constant output voltage, despite widely varying input voltages. The regulated output voltage may be either higher or lower than the unregulated input voltage, but in most applications the output voltage will be lower than the input voltage. The most common type of switching regulator is a three-terminal system that accepts an unregulated DC input voltage and provides a regulated DC output voltage. We will examine this type of regulator first, since it has more universal applications.

Switching regulators used to convert AC line voltage into a DC output voltage are normally implemented by the addition of a conventional rectifier-type power supply to first change the AC into unregulated DC. Switching-type power supplies used to provide a regulated AC output voltage are most often referred to as inverter-type power supplies; these are examined in the next chapter.

A SIMPLE SWITCHING REGULATOR

The simplified diagram of the switching-type regulator shown in Fig. 2-1 will provide basic insights into the various

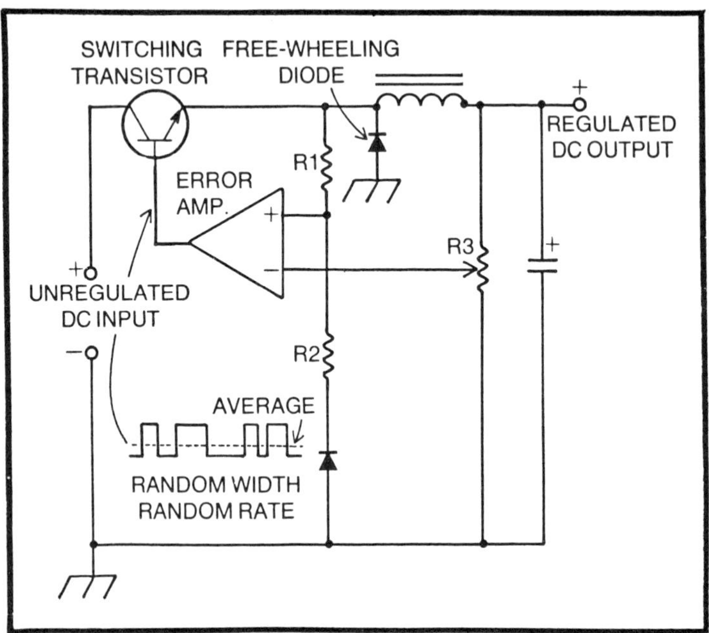

Fig. 2-1. The self-excited switching regulator is self-oscillating. While simple in appearance, the design is complicated by the fact that the oscillatory behavior of the circuit is affected by input and output conditions that vary both the pulse width and pulse rate of the switcher.

techniques for approaching the goal of dissipationless regulation. Obviously, the emphasis in power supply design is primarily on voltage, though sometimes current regulation is also needed. In either case, the operation of the regulator depends upon setting up a feedback loop to monitor the output voltage or current and to provide a correction or *error* signal that is used to automatically adjust the operation of the switching device. This type of feedback operation is able to stabilize the regulator's output, despite variations either in the unregulated input voltage or in output load parameters. Any regulated supply can, for special applications, be divested of its automatic stabilizing operation by simply severing the feedback loop. This may be desirable, for example, in motor-control applications, where manual control of motor speed is sometimes required. However, most electronics are best served by automatic, closed-loop operation; thus, our discussions will focus upon these.

In Fig. 2-1, the basic configuration and main elements of a simple voltage-regulated switcher are shown. In order to center attention on the switching circuitry, the unregulated power supply is not shown. An interesting aspect of this arrangement is that it is *self-oscillatory*. The oscillation frequency or switching rate is primarily determined by the inductance and capacitance of the output filter, but it is also affected by other factors. The output voltage level is proportional to the average value of the *on* times during successive pulses. Thus, the function of the filter is to average out the amplitudes of the switching pulses to produce a constant output voltage whose amplitude is equal to the arithmetic average of the pulses delivered by the transistor switch. Note that the switching waveform is "wild" in that both pulse width and repetition rate are subject to variation. This imposes no great problem, however, for the action of the feedback loop is such that regulation is implemented by variations of pulse width, pulse rate, or of both of these parameters.

In this supply, and in most others, the so-called "free-wheeling" or "catch" diode will be seen. From our discussions of the half-wave rectifier, it will be recalled that a resistance connected in this circuit position enabled the current flow in the load to be more sustained than otherwise, but that the resistance had the objectionable feature of dissipating power. The free-wheeling diode extends the current-sustaining phenomenon into the realm of usefulness, for the diode provides high conduction when it is needed, and negligible dissipation otherwise.

The sense of simplicity that is derived from inspecting this self-oscillatory regulation can be somewhat deceptive. You might well ask, for example, why such a simple arrangement was not popular a decade ago. Part of the answer resides in the circuitry, implied but not shown, within the symbol of the error amplifier. Whether obvious or not, a lot is demanded from this functional element. If you want a lot of performance out of the switching regulator, and at the same time wish to avoid physically massive filter components, a nominal switching rate is needed that is many times the power line frequency.

For reasons to be discussed later, this tends to place the operation in the vicinity of 20 kHz. A cheap error amplifier (or comparator) then becomes an elusive object.

A decade ago, suitable switching transistors, free-wheeling diodes, and even electrolytic filter capacitors were only marginal, insofar as their suitability for high-frequency switching supplies was concerned. And the cost of acceptable components was often prohibitive. The liberties you could take with design simplifications were also limited, because competition with linear regulators, where superb regulation and noise-free operation were obtained relatively easily, made the cost and effort involved in producing a good switching regulator unjustifiable.

One of the really significant turning points in both the technological evolution and the economics of the switching-type supply was the advent of inexpensive, highly reliable integrated circuits. Thus, for a dollar or two you can now acquire an IC operational amplifier (op-amp) for use in the schematic position, depicted by the triangle in Fig. 2-1. Such an op-amp might comprise the equivalent of a dozen or more discrete elements, and these are direct-coupled for DC operation. Even more compelling to the designer than the op-amp is the IC linear regulator, for these contain not only an op-amp but an excellent self-contained voltage reference. The thermal stability and electrical performance of this voltage reference exceeds that generally attainable from simple zener diodes. On a cost basis, it is competitive with voltage-reference diodes and with other circuits such as those using constant-current elements in conjunction with precision resistors. Such voltage regulators, while originally intended for linear operation, are also quite suitable as building blocks in switching regulators. In fact, most manufacturers of linear voltage regulators will gladly provide application notes showing how to use their regulators to make switchers.

THE BEST OF TWO WORLDS

The combination of switching and linear techniques has led to levels of performance not readily attainable using one

28

technique alone. Although switching-type supplies are often simple configurations involving an electronic switching or chopping element, the trend of development increasingly favors a *system approach*, wherein other functional blocks, such as linear regulators, are also involved in the overall scheme.

Consider, for example, a regulated power supply consisting of a switching-type regulator followed by a linear regulator. In such a setup, the switching circuitry can be considered as a *preregulator* for the linear output regulator. This is representative of the simplest of system approaches, but even this system has compelling features. The linear regulator is superior to the switching type in ripple attenuation, in the development of low output impedance, and in obtaining regulation within tight boundaries. But, for many applications, the low operating efficiency of the linear regulator presents severe thermal, electrical, and packaging problems. The low efficiency of the linear regulator is greatly aggravated with increased input voltage from the unregulated power supply. In order to take care of fluctuations in the AC power line, as well as worst-case conditions of load and temperature and a modest margin of safety, the DC input voltage often must be much higher than is compatible with reasonable operating efficiency. Unfortunately, the dissipation in a series-pass transistor increases in proportion to the voltage it must drop to produce its regulated output voltage.

It should be evident that a preregulator would relieve the series-pass transistor of a considerable burden, and that the overall efficiency could be increased if the preregulator itself was very efficient. With such a combination, line voltage could increase greatly with negligible worsening of the overall efficiency. The preregulating switcher might operate at an efficiency of 85%; the linear regulator might operate at a nominal efficiency of 50%, and the switcher would prevent it from dipping below this figure at high line voltage. At high power levels, an improvement of 10–15% in operating efficiency can be quite meaningful. Later, we investigate a number of other switching-type power supplies utilizing the system approach.

THE SELF-EXCITED SWITCHING REGULATOR

The self-oscillating switching regulator in Fig. 2-1 excels in its utter simplicity. While the operation of the switcher is determined in part by the inductance and capacitance of the output filter circuit, its operation is externally affected by both the input supply and output load. As a result, the pulse width and pulse rate of the self-excited regulator may vary considerably because the output filter performs the primary task of filtering instead of tuning.

In the self-excited switcher, the relationship between pulse width and pulse rate is not entirely random, since the switcher is required to deliver a specific output current and voltage to the load. In general, however, the switcher tends to vary its pulse rate in response to changes in load current and to vary its pulse width in response to changes in unregulated DC input voltage. This is a generalization and is not meant to imply that the two actions are entirely interdependent, but rather that these are the dominant effects.

To a certain extent, the design of a self-excited switching regulator is greatly complicated by the fact that *both* pulse rate *and* pulse width are permitted to vary. As a result, the behavior of the self-excited switcher is difficult to predict. In practice, the initial design must be set up using very loose approximations, sufficient to obtain a working circuit; subsequent refinements are then made during actual operation, at which time changes in component values of up to 100% are not at all uncommon.

The operation of the self-excited switcher in Fig. 2-1 is easily explained in general terms. The sense amplifier performs a *comparator* function; that is, it compares the output voltage of the switcher (fed back via resistor R_3) to a reference voltage, which in this case is provided by a zener diode. When the output voltage of the switcher drops too low, the sense amplifier turns on the switching transistor; when the output voltage rises too high, the switching transistor is turned off. The only problem encountered in this simple scheme is in making the switcher oscillate reliably under *all* conditions of DC input voltage and output load current.

To insure reliable operation, it is necessary to add a certain amount of hysteresis (provided by R_2 and R_3) into the feedback loop to introduce sufficient positive feedback for guaranteed oscillation and fast switching of the transistor. But the presence of these resistors in the feedback loop also affects the pulse width, pulse rate, output AC ripple voltage, and average DC output. Consequently, while the self-excited switcher can boast of having few components, the design of the switcher is actually complicated by the fact that there are *not enough* components to isolate the controlling parameters that govern its operation. Still, the self-excited switcher remains very popular in less-demanding applications where its low cost and few parts are very compelling features.

THE EXTERNALLY EXCITED SWITCHER

In the self-excited switcher of Fig. 2-1, it was observed that both the pulse width and pulse rate varied freely, and that such freedom introduced design problems in guaranteeing reliable oscillation. At the sacrifice of a slightly more complex circuit, it is possible to establish two possible modes of operation, simply by making either the pulse width or pulse rate a constant parameter. These two modes can be summarized as follows:

- Regulation is obtained by holding the pulse rate (frequency) of the switcher constant and permitting only the pulse width (the *on* time) to vary.
- Regulation is obtained by holding the pulse width constant and permitting only the pulse rate to vary.

In the self-excited switching regulator, the division of labor between varying the frequency and varying the pulse width is more or less capricious. No doubt, one parameter or the other could be made relatively constant, thereby shifting the corrective action to the parameter with greater freedom of deviation. These objectives are more easily achieved in switching circuits which do not rely upon self-oscillation for their basic operation, but instead utilize separate oscillators to produce the waveform which turns the switching transistor on or off. It then becomes easier to independently modulate or control either rate or duration without affecting the other.

It is often desirable that circuit function blocks be isolated to permit them to be optimized for one exclusive function. Otherwise, we find ourselves in essentially the same position as the early proponents of "reflex" radio sets. In these now obsolete circuits, one stage or channel might handle RF. IF, and perhaps audio frequencies as well. While the idea is intriguing, there are practical drawbacks in attempting to optimize all of these operating modes at the same time. In the case of the regulator, more reliable performance and better operating characteristics are generally achieved by allowing the switching transistor to be simply a voltage-controlled switch. We can also separate the functions of positive and negative feedback and isolate the reference voltage at the sense amplifier. (By dispensing with the need for self-oscillation, the positive feedback path is entirely eliminated).

With separate oscillators, you are free from such malfunctions as failure to start, say at low temperatures and with a heavy load, and from certain transient disturbances which tend to "shock" the LC filter into oscillation at a "natural" frequency unsuitable for regulation. Certain changes, such as simultaneous line and load changes, can also drive the pulse rate and pulse width in antagonistic directions, thereby upsetting the regulation. Finally, as a designer you can more accurately specify the values of the filter components if only one switching parameter is allowed to vary. Switching regulators of this type are easier to mass produce within given specification boundaries. The alternative—overdesign or the "brute-force" approach—can easily negate the apparent low-cost feature of the self-excited switcher.

The Fixed Frequency, Variable Pulse-Width Regulator

The operation of the system in Fig. 2-2 is similar to that of the self-excited type, but the switching rate is now fixed at the frequency of the separate oscillator. From a practical point of view, the operation of the supply can now be optimized by merely shifting the frequency until optimum results are obtained. This is often desirable because the initial design cannot take into account such variables and unknowns as the

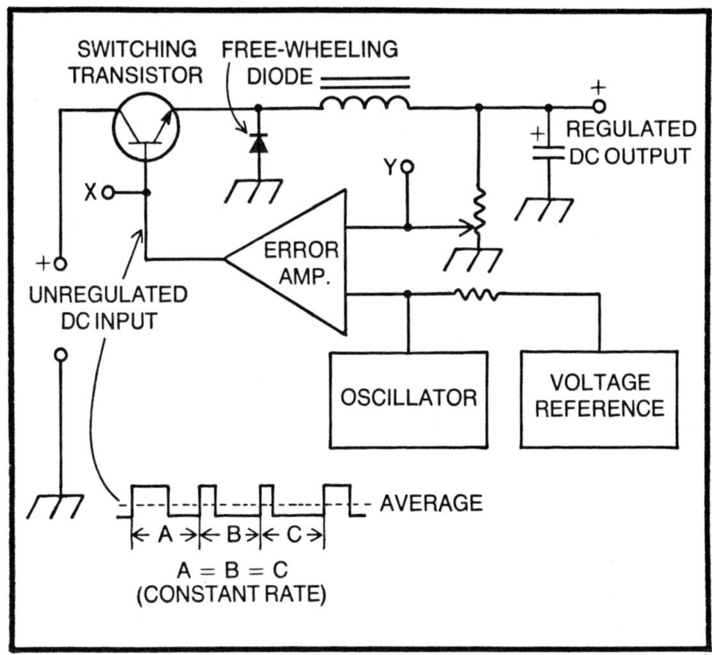

Fig. 2-2. Fixed-frequency switcher. In this arrangement, the pulse rate or frequency of switching is established by a separate oscillator. Regulation is then accomplished by permitting the pulse width or ON time to vary, a technique often referred to as duty-cycle or pulse-width modulation.

operating inductance and Q of the choke, the inductance and effective series resistance of the filter capacitor, and the specifications of the switching transistor and the free-wheeling diode. It is also nice to have independent control of the frequency from the viewpoint of the powered equipment; switching regulators generally produce greater ripple than their linear counterparts, and it is often profitable to have the option of adjusting the ripple frequency to minimize load disturbances.

Constant Pulse Width, Variable Frequency Regulator

The alternate fixed parameter scheme is shown in Fig. 2-3. Here, the pulse rate of the monostable or "single-shot" multivibrator is varied by trigger pulses obtained from a voltage-controlled oscillator. The duration of the *on* pulses applied to the switching transistor are determined by the time constants within the multivibrator, and remain fixed. This is a

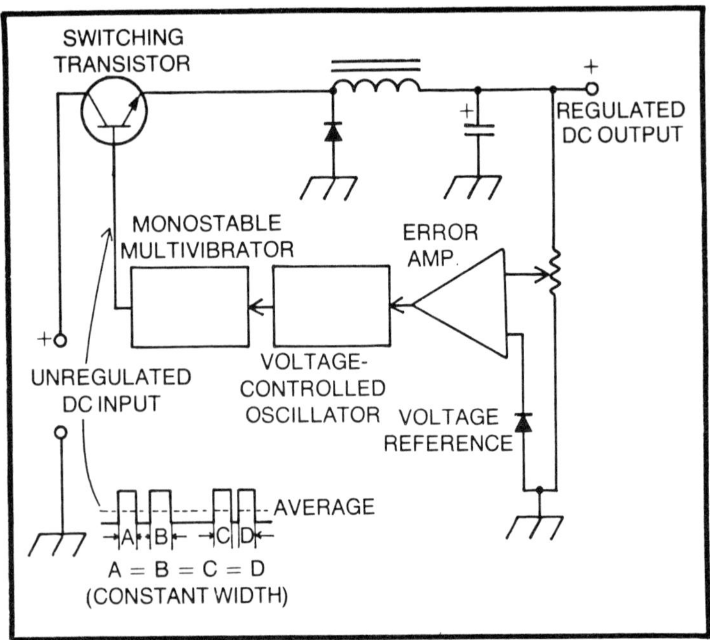

Fig. 2-3. A fixed width is obtained using a monostable multivibrator, which, when triggered, emits a pulse of fixed time duration. Hence, the on time of the switching transistor is constant, and regulation can be achieved only by variation of the pulse rate, or frequency.

rather elegant scheme that allows independent adjustment of both the pulse width and the pulse rate. It is not easy to say that either this method or the one previously described employing fixed frequency is better, although it is possible to cite certain advantages for one or the other. Any possible superiority is a function of the design method, so that it is probably best to say that while one designer could attain superb results with one approach, another designer might achieve equally favorable results with another. There are fine points of difference having to do with output ripple, allowable adjustment range of the DC output voltage, and isolation from line transients, but both fixed-parameter techniques have good performance capabilities.

The Shunt Switcher

Thus far, all of the switching regulators discussed have utilized a *series* switching element. We know from linear regulators that *shunt* elements are sometimes advantageously used, so it would be only natural to use a shunt-connected switching element to accomplish regulation. It so happens that the shunt switcher is practical, and would assume the form depicted in Fig. 2-4. This particular arrangement uses a constant-pulse-width technique, but fixed-frequency schemes can also be implemented. The shunt switcher develops output voltage levels which are *higher* than that of the unregulated input supply.

A possible advantage of this switching technique is that it handles radio frequency and electromagnetic interference easier, since the current through the switching transistor following turnon is a ramp, rather than a step function. A disadvantage of the shunt switcher, however, is that it has a lower efficiency than the a series switcher; this results primarily from the fact that the input voltage is less than the output voltage, so that voltage drops occurring in the input circuitry waste much more power. But the fact that the shunt

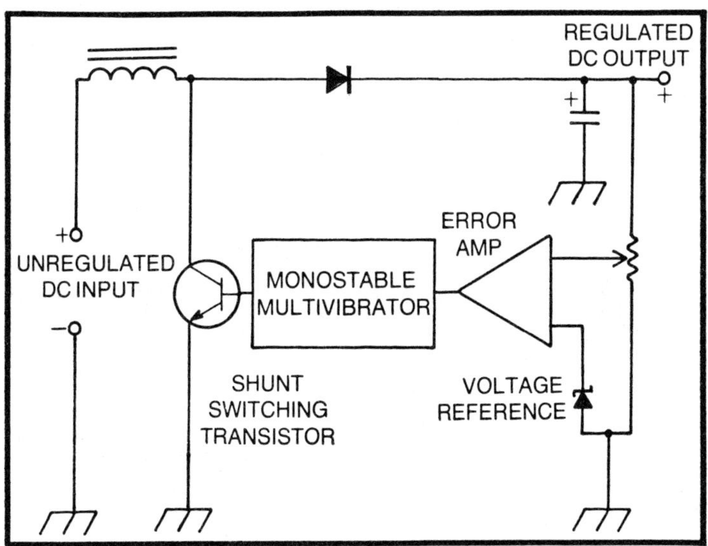

Fig. 2-4. The shunt switcher provides a simple method for obtaining a much higher output voltage when only a low-voltage DC source is available, as in electronic watches and calculators. While the shunt switcher may take many forms, it is not generally suited to self-excited operation, since the on time of the switching transistor must be carefully limited to prevent saturation of the inductor and excessive transistor currents.

switcher is able to produce a higher output voltage makes the shunt switcher very useful in converting low-voltage DC battery power into higher voltages.

The shunt-switching technique offers an inherent protective feature in the event that the switching transistor locks up in either an open or closed state: in either case, the load would be spared the likelihood of destruction. If the switching transistor stays in its *on* state, it would act as a "crowbar" short across the supply, and with appropriate design could interrupt a fuse or circuit breaker—even destruction of the switching transistor is generally preferable to damaging costly equipment. If the switching transistor remains *off*, there would be a decrease in the output voltage, and the load would again be protected from damage, since the output voltage of the regulator would then be no greater than the DC input voltage.

The operation of the shunt switch is analogous to that of other circuits using the "flyback" principle. Included among these are automobile ignition systems, electronic switching supplies for geiger counters, and high-voltage supplies in television receivers. The difference in the shunt switching regulator rests in the use of a feedback arrangement to stabilize the output; otherwise, all flyback circuits make use of the counter electromotive force developed by either an inductor or the primary winding of a transformer.

OBTAINING VARIABLE PULSE WIDTHS

The block diagram of Fig. 2-5 represents one method of achieving the constant-frequency mode of operation depicted in Fig. 2-2. Only the control circuitry is shown, but its association with the switching transistor and accompanying components is indicated by reference to points X and Y in both figures.

At first glance, the arrangement of Fig. 2-5 might appear rather complicated; however, the operating principle is straightforward, and the use of integrated circuits makes for a low component count. This is an excellent example of translating concept into hardware, a task that would be severely hampered if each functional block were constructed

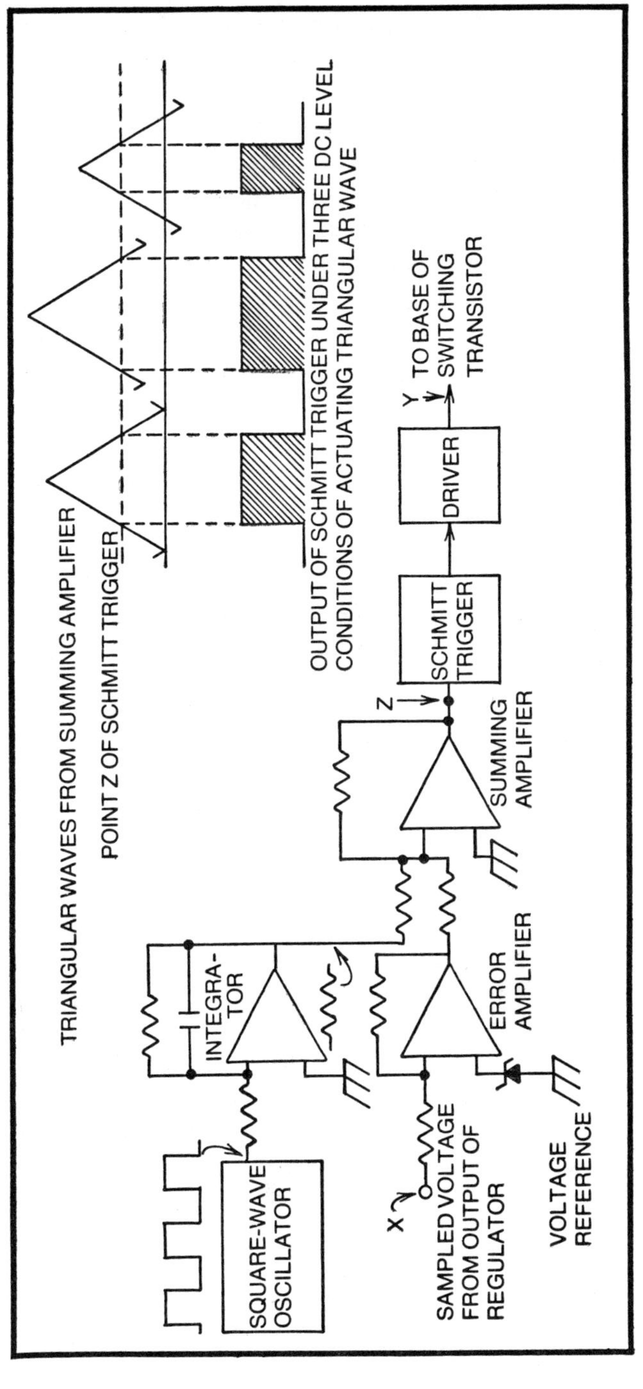

Fig. 2-5. A pulse-width modulator. This circuit utilizes a Schmitt trigger to perform a voltage comparison. In practice, the Schmitt trigger would incorporate a small hysteresis, so that its turnon voltage would be slightly higher than its turnoff voltage, but for simplicity this is not illustrated.

from discrete devices. Here, just a few inexpensive ICs together with some external components produce a complete pulse-width modulator.

The square-wave oscillator does not have its frequency affected directly, so that it is possible to vary the frequency manually to optimize the operation of the circuit. A simple square-wave generator could be constructed using a multivibrator, since the accuracy of the frequency is often critical. The integrator converts the square wave into a triangular wave, which is then summed with the output voltage from the sense amplifier. The Schmitt trigger is a level-sensitive circuit that produces a rectangular output pulse, as illustrated in the figure. As the output voltage of the sense amplifier varies in response to the output of the switching regulator, the triangular waveform is shifted up and down, varying the trip point of the Schmitt trigger. This produces an output pulse of varying duration that is then applied to the switching transistor. A *driver* block is shown in Fig. 2-5, representing an extra stage of amplification, but this

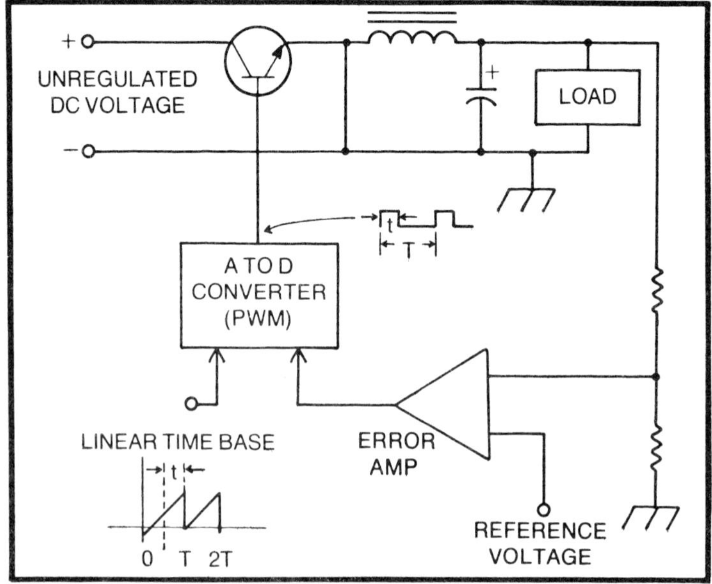

Fig. 2-6. Generalized pulse-width modulator. In its simplest form, a switching regulator requires only a ramp-type input voltage, a reference voltage, and a few integrated circuits to make it a reality.

is frequently omitted in switching regulators of one-ampere capacity or less.

There are, of course, many other ways to implement the same functional mode of operation, including purely digital techniques. For precise frequency control, an external frequency source could be used. And there is no requirement that the waveform be square or even triangular wave; many systems incorporate ramp-type waveforms and sine waves. An example of the linear-ramp technique is illustrated in the block diagram of Fig. 2-6, and typifies many such designs.

The functional-box approach of the switching regulator system in Fig. 2-6 utilizes a block marked *analog-to-digital converter*. Such boxes already do exist in integrated circuit and modular form, but the important fact is that the box generalizes the basic function to be achieved in obtaining a variable pulse-width output. Reduced to its simplest form, the switching regulator no longer represents a stumbling block to the designer.

3
Switching-Type Power Supplies

The simple switching regulator presented in Chapter 2 represents one approach to efficient power supply design. In this chapter, several other power supply systems will be presented, many of which utilize the simple switching regulator as a part of their complete system. Since the simple switcher has already been discussed, its function in a larger system will be readily understood as will variations of the simple switcher, using SCRs and triacs instead of transistor switching elements.

AC-POWERED SWITCHERS

Switching-type power supplies that use arrangements such as those shown in Fig. 3-1 represent a quantum jump in power supply advancement. These regulators are, in essence, systems of functional blocks. The major feature of these regulators is the elimination of physically massive power transformers and other power-line magnetics. The net result is a smaller, lighter package and reduced manufacturing cost, resulting primarily from the elimination of the 60 Hz components.

In both systems in Fig. 3-1, the incoming AC power is immediately rectified and filtered . This unregulated DC power

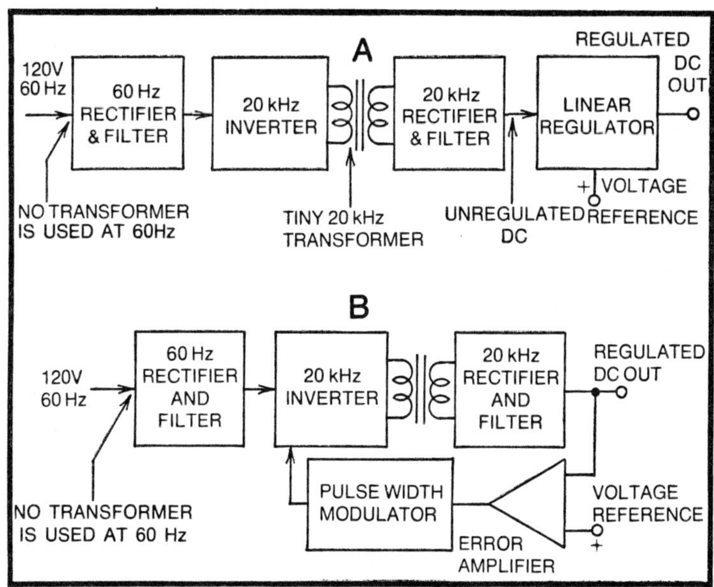

Fig. 3-1. Two approaches to switching-type power supplies. The hybrid approach in A combines the high efficiency of a switching inverter with the precise voltage control of a linear regulator. The switching regulator in B provides even greater efficiency, with some sacrifice in output ripple voltage, by using only switching techniques to perform regulation.

is then converted to a frequency many times that of the power line, often in the 20 kHz range. Then, by one or another type of regulator, the high-frequency power is changed to regulated direct current. Because of the high switching frequency, magnetic core and filter components are minuscule compared to their 60 Hz counterparts. Moreover, the 60 Hz filter at the front end of the system can be quite simple, since subsequent regulation provides electronic filtering power-line ripple frequencies.

Schemes of this kind represent a bold new concept in power supply design. There are actually many versions, all incorporating the basic idea of *internal frequency setup*. The generation of the high-frequency power is generally performed by a saturable-core inverter, although other nonsaturating types are used. A large-scale implementation of this design philosophy was carried out by Tektronix Inc. when the firm introduced a new line of oscilloscopes which, though brimming with the latest test functions, were dramatically more compact than previous instruments. The use of the sophisticated switching supplies enabled the oscilloscopes to operate at reduced temperature and consume appreciably less power than their predecessors.

VOLTAGE-STABILIZED INVERTER

Figure 3-2 provides a more detailed inspection of one way in which the frequency setup can be accomplished in the inverter block of Fig. 3-1B. The inverter depicted is *not* the self-excited type common to such applications as the generation of AC from automotive batteries; rather, this inverter is *driven* from a logic-controlled source, starting with a square-wave oscillator. The two switching transistors function essentially as a push-pull amplifier: each transistor is alternately driven full *on* and then *off*, and a rectangular wave is developed across the primary of transformer T_1. Unlike self-excited inverters, the operating frequency of this inverter is not dependent upon saturation of the core of T_1.

The prime operating feature of this arrangement is that although the square waves originate from the oscillator, across transformer T_1 the duration of the rectangular wave is varied by the logic circuitry in the *pulse-width control* block. This variation occurs in response to the amplified error signal in such a way as to regulate the level of the DC output voltage. When the output must be increased, the pulses of the rectangular wave are longer; the converse is true when the output voltage must be decreased. An inverter of this type has excellent performance and is virtually devoid of the malfunctions which often plague self-excited types; for example, there are no starting difficulties, the likelihood of the circuit bursting into oscillation is greatly reduced, and the pulse-width control insures that both switching transistors will never be simultaneously *on*. A *phase splitter* provides a balanced push-pull signal to the two processing channels; it can be simply a transformer with a centertapped secondary, or it can be an active circuit designed to perform the same function.

Unlike most of the switching-type regulators previously shown, neither a filter choke nor a free-wheeling diode is

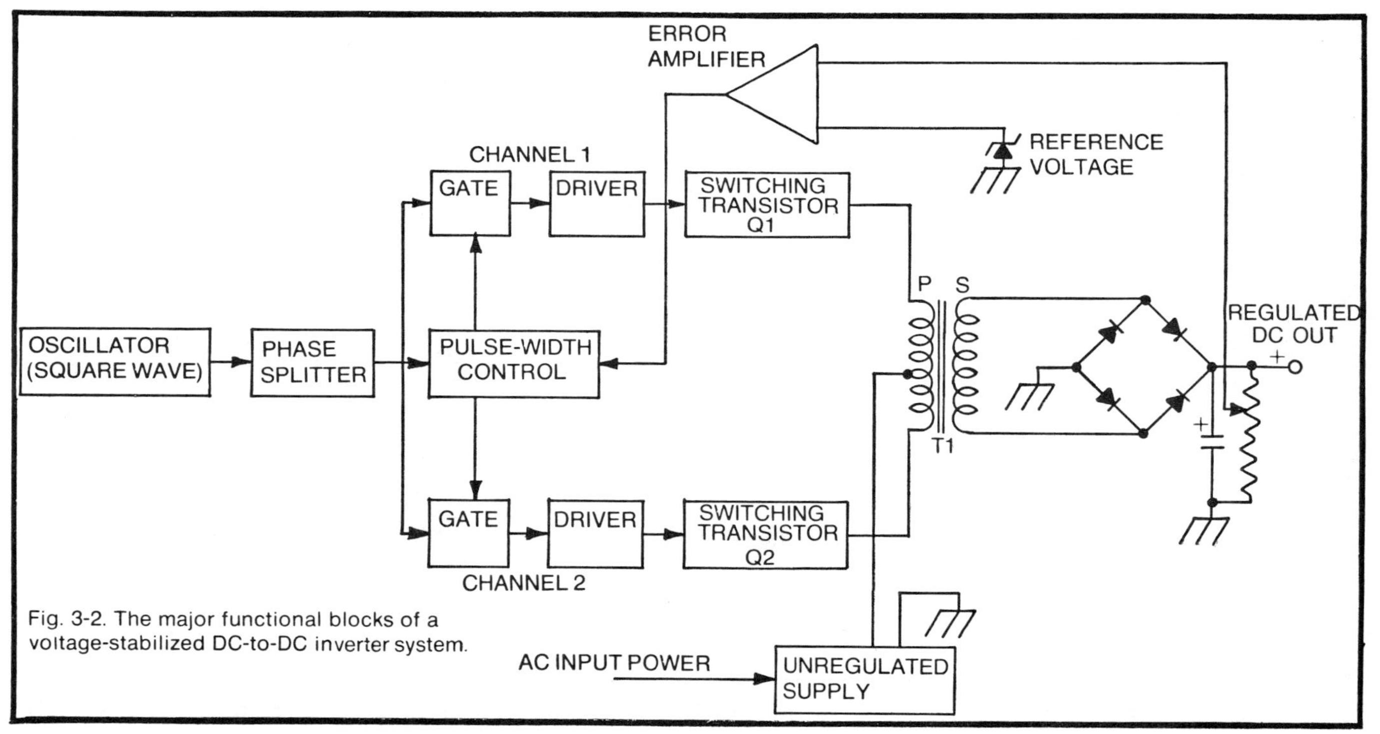

Fig. 3-2. The major functional blocks of a voltage-stabilized DC-to-DC inverter system.

necessary with inverters. One of the objectives in the design is to use such a high switching frequency that good filtering of the output can be accomplished with a relatively small capacitor. These considerations also tend to reduce feedback loop instabilities when using high-gain error amplifiers.

SCR FULL-WAVE REGULATOR

The circuit arrangement shown in Fig. 3-3 represents a practical way of combining full-wave rectification with the desirable features of SCRs. This technique is easy to implement, and requires neither a centertapped transformer nor critical firing circuitry for the gates of the SCRs. Regulation is achieved through phase control of the SCRs, but there are some subtle aspects to this circuitry that may not be immediately apparent.

The main rectifying bridge is comprised of two conventional rectifying diodes and two SCRs. This bridge provides no output except when one or the other SCR is triggered into its conductive state. Although the gates of *both* SCRs receive the same turnon trigger, only that SCR which then has its anode polarized positively with respect to its cathode will actually be turned on.

The average current or voltage delivered by the rectifying bridge is governed by the time delay in a given half-cycle until the "favored" SCR turns on; thus, this rectifying bridge will deliver various *fractions* of half-cycles. The greater the fractional part of a full half-cycle, the greater will be the current or voltage delivered to the filter. Inasmuch as the main rectifying bridge is controlled by timing the triggers to the SCRs, it follows that the remainder of the circuitry is largely devoted to the production and timing of the SCR trigger pulses. These triggers are generated by unijunction transistor Q_1, which is connected as a relaxation oscillator; the oscillation frequency is, among other factors, determined by timing capacitor C, timing resistance R, and the effective resistance of transistor Q_2. The occurrence of trigger pulses is controlled by varying the conduction of transistor Q_2. In turn, Q_2 is controlled by the error signal obtained from the error

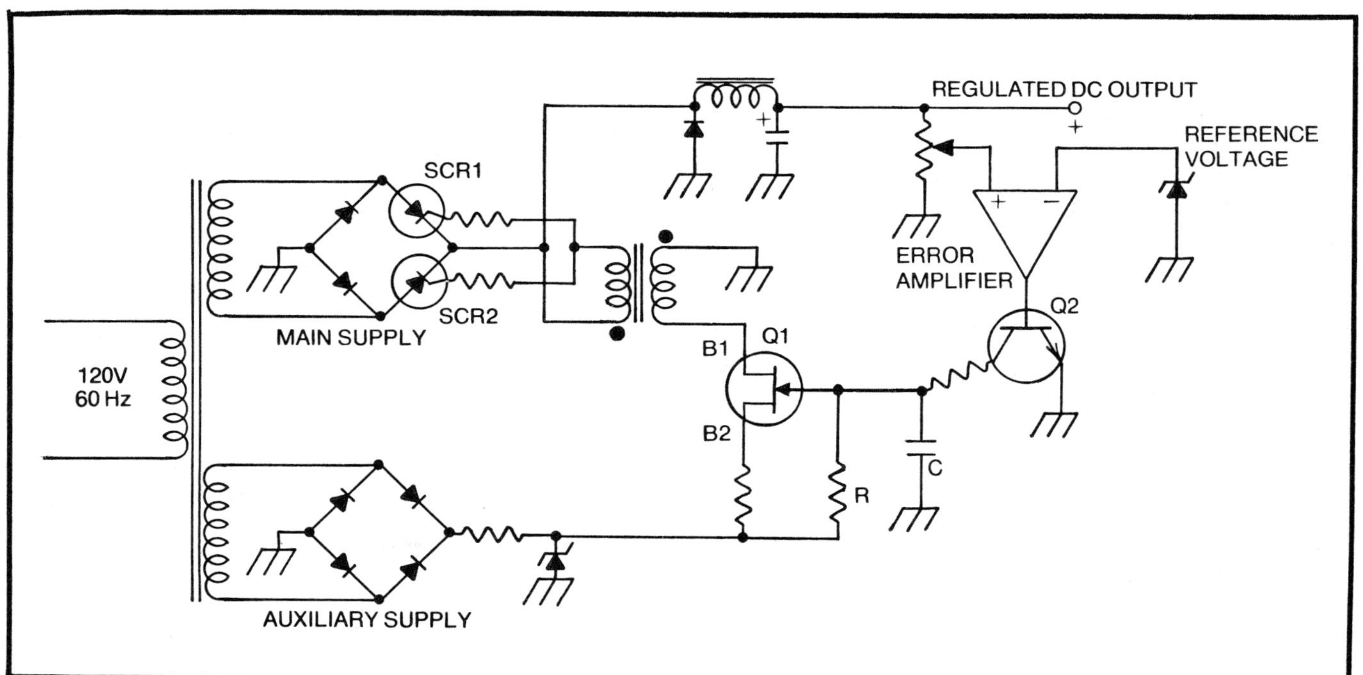

Fig. 3-3. Full-wave switching regulator using SCR switching elements.

amplifier, which samples the DC output level of the regulator, as is the case with most regulating feedback networks.

A method is also needed for synchronizing the UJT oscillator to each zero crossing of the AC wave on the power line. Otherwise, the control of the SCRs will occur in a hit-and-miss fashion. The auxiliary supply is not filtered: there is no filter capacitor following its bridge rectifier. Because of this omission, the ripple in the output causes a momentary dip to zero, twice per cycle of the power line frequency (the ripple frequency is double the frequency of the rectified AC wave). In this case, the output of the auxiliary rectifier dips to zero 120 times per second. Moreover, these dips take place at the zero crossings of the AC line. These dips interrupt the operation of the unijunction circuit, so the relaxation oscillator is not free-running, but is forcibly restarted at each zero crossing. Timing capacitor C commences its charging cycle each time the AC wave crosses zero; each of these charging cycles begins its rise from a near-zero voltage level. When the timing capacitor begins its charging cycle, nothing happens until the voltage developed across its terminals is sufficient to fire the unijunction transistor. When the unijunction transistor fires, a turnon pulse is delivered through the pulse transformer to the SCRs. As previously explained, the favored SCR is then turned on. The same thing then happens during the next half-cycle, only this time the *alternate* SCR is turned on. The SCRs automatically turn off as their anode–cathode voltage declines to zero. The zener diode associated with the auxiliary power supply provides a waveshaping function, making the ripple easier to use for the unijunction timing. It does not enter into the basic philosophy of the system; the line-synchronized dip to zero remains the important aspect of operation.

This type of switching regulator generally merits serious consideration when dealing with loads requiring tens or hundreds of amperes; even high-current capabilities are feasible. The response time of such line-frequency switchers is limited, but that often is not a deterrent when powering heavy loads. (Better performance and economics can be obtained for 400 Hz operation.) It should be mentioned that this scheme enables such easy attainment of full-wave operation that there is hardly any justification for regulators designed around single SCRs in half-wave configurations.

TRIAC SWITCHING REGULATORS

The triac is a more desirable switching element, if considerations of power and frequency allow its use in place of SCRs. In essence, the triac functions as a pair of oppositely poled SCRs connected in parallel. Thus, one has a push-pull or full-wave control technique. The analogy is not exact however, because the triac has a *single* control gate, which leads to both design and operational simplicity. A switching-type regulator employing a triac is shown in Fig. 3-4. Note the symmetry of the illustrated phase-controlled wave. A supply of this kind has relatively slow response, which is inherent in circuits using line-frequency switching. Where this is not a deterrent, such a supply has much to commend it, for triacs are inexpensive and very efficient switching devices.

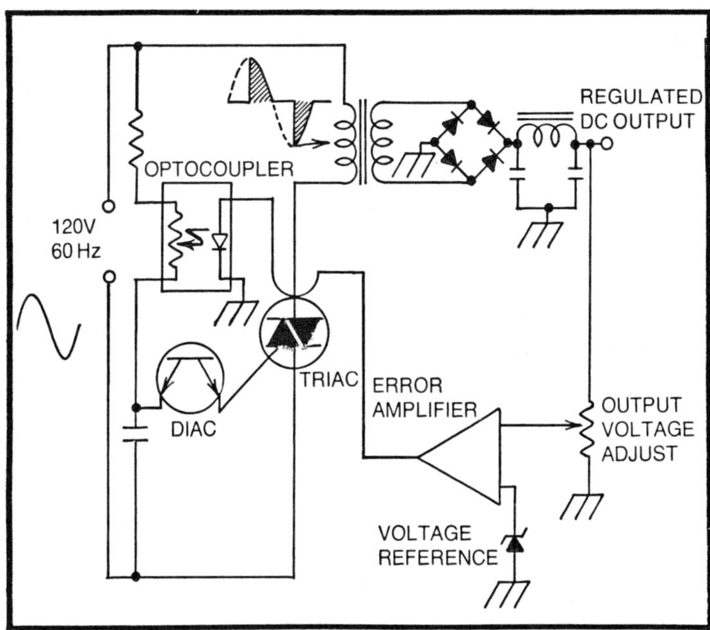

Fig. 3-4. A full-wave switching regulator utilizing a triac as the switching element. Note the use of an optocoupler to provide variable-resistance control over the triac, while at the same time providing electrical isolation between the AC power line and error-amplifier circuitry.

The use of the optocoupler in the feedback loop represents a circuit technique which is growing in popularity. The optocoupler provides electrical isolation, while at the same time transmitting signal information. This avoids problems having to do with DC level translation in the original circuit design, and the circuitry is safer from damage caused by inadvertent grounding or shorting, as might occur from alignment procedures or from certain load conditions. Otherwise, this switching-type regulator involves no new techniques that have not been extensively used in previous thyristor control. The error amplifier functions as a straightforward differential amplifier; it provides linear control and does not saturate, as in some of the previously described regulators. There is no reason to manufacture a switching waveform here, as it already exists in the power line.

Where 400 Hz power is available, this type of regulating supply presents some compelling features. The transformer and filter components would be appreciably reduced in size from their 60 Hz counterparts, and newer triacs may perform reasonably well at this frequency. At present, SCRs are strides ahead of triacs in both power and frequency capability. On the other hand, most electronic applications do not require the high power levels required by traction motors or industrial machinery.

MAKING SWITCHERS OUT OF LINEAR REGULATORS

Those who have constructed *linear* regulators will recall that one of the most commonly encountered malfunctions in these circuits is self-oscillation. This is particularly true when, for the sake of tight regulation, the error amplifier has very high gain. The oscillation may be superimposed upon the DC output level, so that it may not actually interfere with the operation of either the power supply or the load. Generally, however, such oscillation produces all manners of troubles and component failures. Indeed, the oscillation is frequently in the form of a square wave, because the error amplifier alternately saturates at one extreme then the other. At the same time, the "linear" series-pass transistor is being switched between cutoff and saturation.

48

The operational mode of this malperforming linear regulator simulates, in some respects, the intentional operation of the self-excited switcher. In fact, the transformation from linear to switching is often as simple as adding a few extra components. Manufacturers of on-the-chip linear regulators can also provide application notes on switching-type regulators. This is true even if the switching regulator was far from their thoughts during the development of the IC linear regulators: IC linear regulators suitable for use in switching-type regulators are the 723, LM104, LM105, and 550.

The close but long-elusive relationship between a linear and a switching-type regulator is clearly illustrated by the two discrete power supplies shown in Fig. 3-5. In this case, a

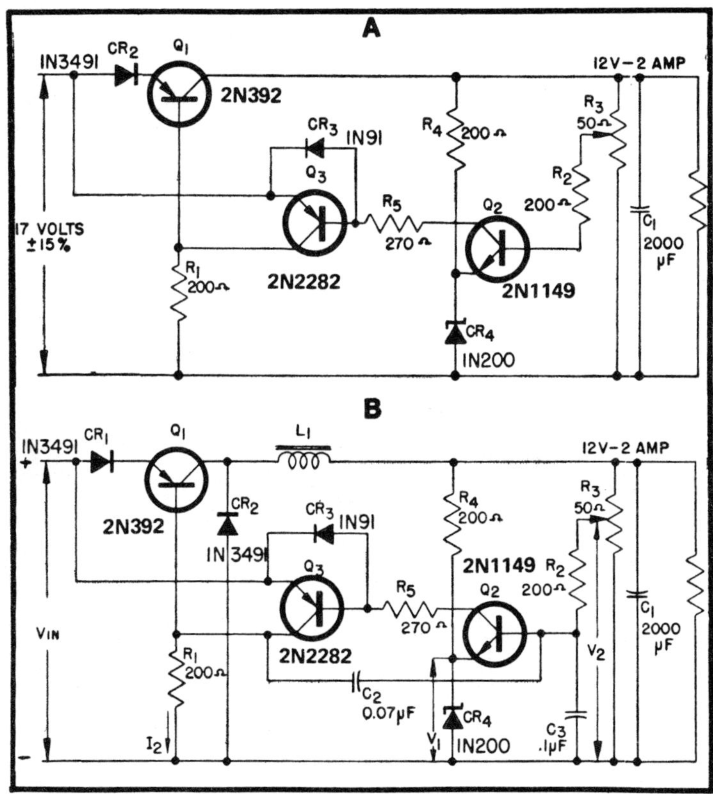

Fig. 3-5. Simple conversion of dissipative linear power supply (A) into high-efficiency switching supply (B). (Courtesy Delco Electronics)

deliberate conversion of the linear circuit was made in order to obtain switching-type regulation. The change is effected by the addition of inductor L_1, the 1N3491 free-wheeling diode, and capacitor C_2 (which converts stages Q_2 and Q_3 into a variable-duration multivibrator).

CONSTANT-CURRENT REGULATION

Although the application demand for voltage-regulated switchers has predominated, it should be appreciated that *constant current* switchers are also feasible. The modifications necessary to accomplish this mode of operation are analogous to the conversion of a linear voltage regulator to a current regulator. In both the linear and the switching-type regulator, current regulation is implemented by sensing the voltage drop developed across a tiny resistance inserted *in series* with the load rather than by sampling the voltage produced *across* the load, as in voltage regulators. When a voltage regulator maintains constant voltage across this series resistance, it thereby regulates the current flowing through the load. Thus, the basic action of the two modes of regulation is quite similar.

From a dynamic standpoint, there are some important differences. In voltage regulation, one strives for as *low* an output impedance as possible; the opposite is true for current regulation. Ideally, the equivalent circuit of a perfect current regulator incorporates a generator with *infinite* impedance, but in a practical current regulator, a high impedance will suffice, provided that it rises with frequency. Thus, the residual inductance in the filter capacitor and its leads is not as detrimental here as in voltage regulators. As a matter of fact, it is often desirable to greatly *reduce* the value of the filter capacitor in current regulators to achieve better transient response. This may cause instability in the feedback loop, but corrective measures can usually be applied elsewhere in the circuit; for example, at the frequency compensation terminal of the IC regulator. Much depends upon the nature of the application of the current regulator; for battery charging and electrochemical processing, rapid response is not needed, and minimal difficulty should be

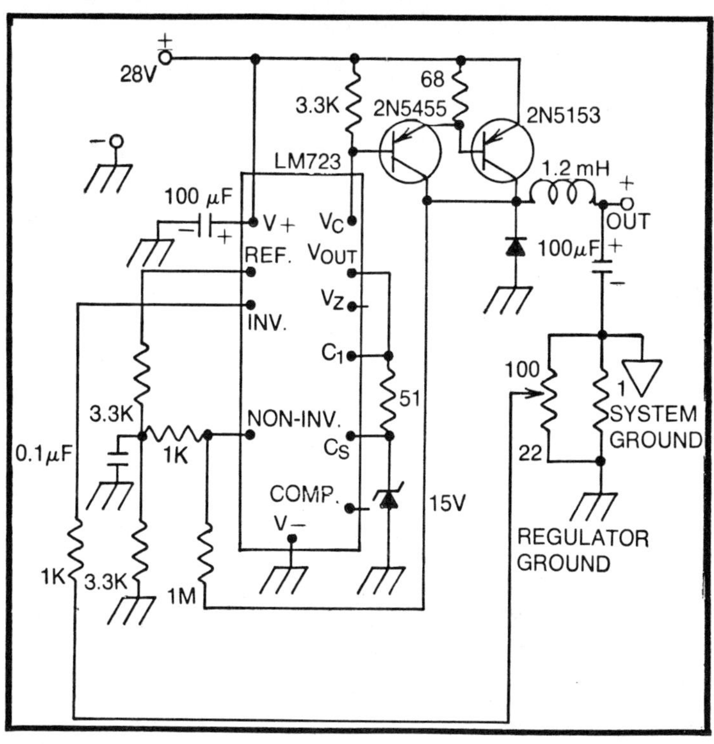

Fig. 3-6. Example of switching-type current regulator. Note that the regulator and external load should not share a common grounding system.

encountered in altering conventional voltage-regulating circuits to perform current regulation.

In Fig. 3-6 an adaptation of the popular LM723 is shown. Just as with linear current regulators, one must be very careful regarding *grounding* arrangements. A straightforward way to circumvent ground conflicts is to allow the unregulated DC supply, together with the regulator circuit, to "float." The standard ground symbol in Fig. 3-6 would then signify nothing more than a common interconnecting bus, but one *isolated* from the *true* ground at the powered equipment.

A further means of protection for the constant-current supply is often obtained by inserting a resistor or diode in series with the load. In this way, active loads such as batteries and inductive windings are prevented from unleashing energy back into the regulator circuitry.

4
Noise?
What Noise?

Not too many years ago, engineers declared that the switcher was doomed to a shadowy and nebulous existence in electronic systems. The dire forecast was predicated upon the "fact" that switching supplies were inherently noise generators. Overlooked by these prophets of doom were matters such as the following:

- Any time a rectifying circuit is associated with a large input capacitor, highly nonsinusoidal current waveforms are produced. Because of the power levels often involved, strong harmonics are generated over an extremely wide frequency spectrum. Thus, the average linear regulator is not so innocent when it comes to noise generation. Typical current waveforms in capacitor-input rectifier circuits are shown in Fig. 4-1.
- The noise production from a switching-type power supply should always be compared with that of a linear supply having the *same* power rating rather than to an ideal, noiseless power source. When such a comparison is made, it will often happen that the switcher is actually the smallest noise contributor.
- It often happens that digital *loads* generate more noise than switching-type supply itself.

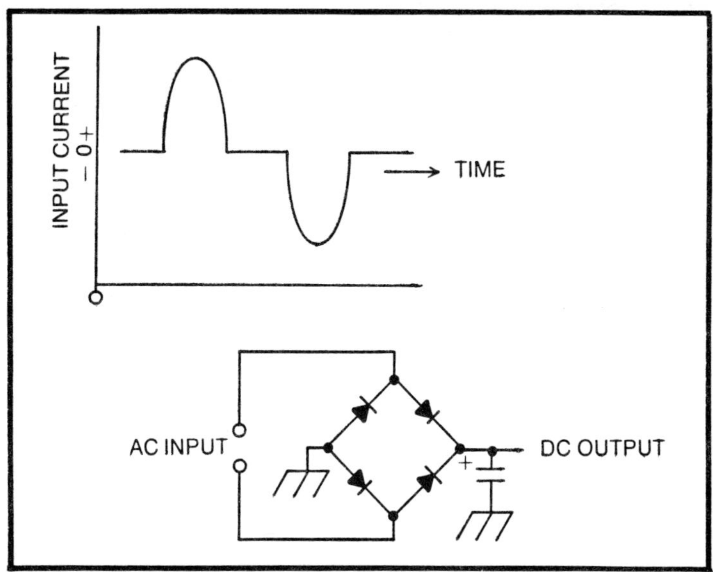

Fig. 4-1. Harmonic-rich current waveform at input of capacitor-loaded rectifier. The pulse-like wave generates harmonics. Although the power-line
frequency is 60 Hz, a spray of odd-order harmonics, ranging deeply into
the megahertz region, can result from such a waveform.

- Many powered systems utilize CMOS or other logic with high noise immunity.
- Considerable understanding has been attained with regard to noise filtering and suppression techniques. Even though the switching process may be prolific in the generation of harmonic and spurious frequency components, this energy can easily be contained within the physical confines of the power supply.
- There is often the option of choosing the switching frequency, so that the residual noise imposes minimal interference with the particular equipment being powered.
- Some switchers can provide greater rejection of power-line transients than linear regulators.
- Because of new ICs, fast-recovery diodes, superior capacitors, and other improved components, the overall operation of switching regulators is better than ever. This enables more effective tradeoffs to be made for the sake of noise reduction.

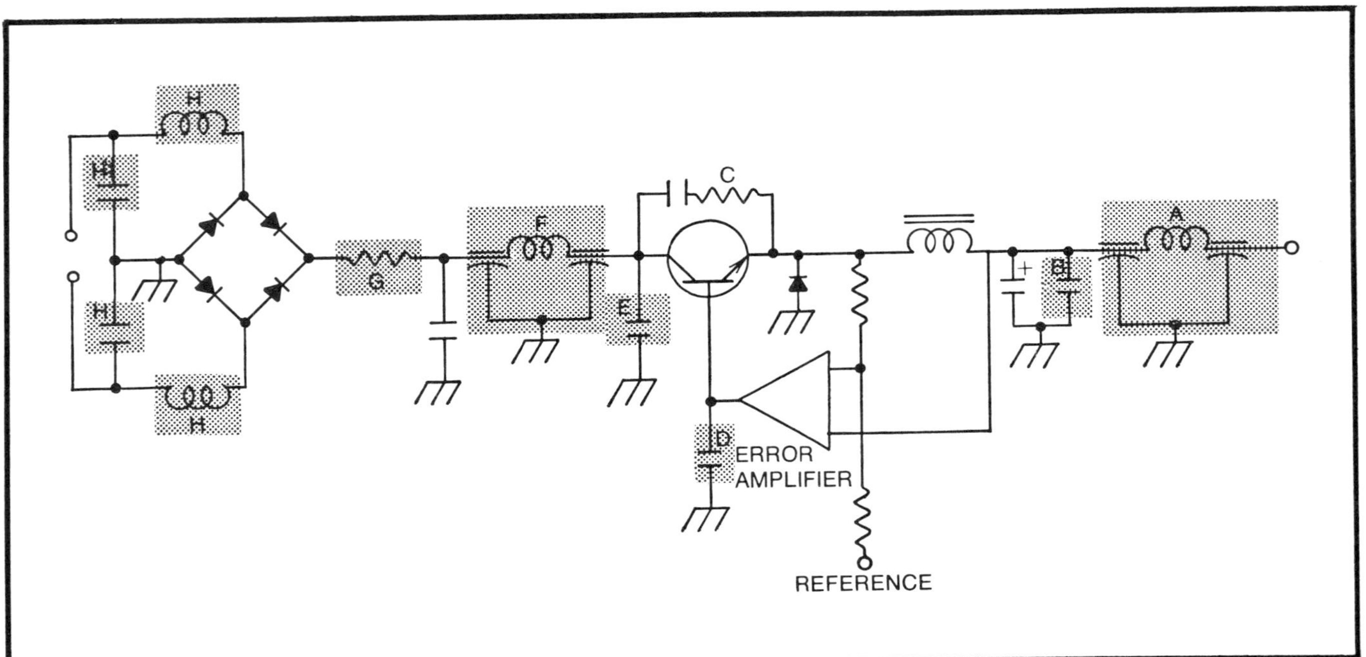

Fig. 4-2. Techniques for reducing electrical noise in a switching supply. The shaded components provide slowdown, filtering, and isolation functions, as described in the text.

So much for the defense of switchers. The fact remains that they are *potentially* offensive with regard to noise production. Admittedly, there are some applications where it would not be wise to consider the use of these supplies. For example, a sensitive radio receiver is best powered by linear-type supplies. Similarly, instrumentation which is predicated upon a high signal-to-noise ratio could have its performance degraded by a switching-type power supply—particularly when the signals being monitored or detected are extremely weak.

For a great many applications, success will depend much upon design and installation techniques. Although one must be guided by logic and by available facts in the implementation of shielding and grounding methods, the final optimization is invariably empirical. One must be willing to experiment, to observe, and to try various combinations and permutations. The reward is often a surprising reduction in conducted or radiated noise. Unlike many electronic products, the switcher experimentation does not end with the breadboard, but is continuous in the initial prototype models. Finally, after various modifications, noise will have been reduced to the point of diminishing returns, with respect to the expenditure of man-hours and money. At this stage, reasons can probably be found to account for some of the strange noise-reduction methods.

NOISE-REDUCTION TECHNIQUES

Figure 4-2 shows a number of commonly employed noise-reduction techniques. Most of these adversely affect efficiency, regulation, or other performance parameters, so they must be implemented judiciously, with due regard for permissible tradeoffs.

- Output filter *A* can be used to attenuate high-frequency noise in the RF region, as well as ripple at the switching rate. This filter is *outside* of the feedback loop; therefore, the DC resistance of the inductor will degrade the voltage regulation. As shown, this filter is a *feedthrough* type. The "inductor" is simply a small segment of the conductor surrounded by a ferrite bead, and the "capacitors" consist of the built-in ceramic insulation. Feedthrough filters are excellent for the removal of high-frequency noise, but are limited by their relatively low inductance for attenuation of lower-frequency noise components. Conventional filters are usually necessary for filtering out switching frequencies and lower-frequency transients. In order to recover the lost regulation caused by the additional output filter, attempts are sometimes made to include it within the feedback loop. If this can be done without upsetting the loop stability of the regulator, it is desirable, but more often than not, this will lead to erratic operation.

- Capacitor B is often a tenth to a hundredth the value of the output capacitor to which it is connected in parallel. This added capacitor provides bypass action for the higher frequencies where, because of internal resistance and inductance, the main filtering capacitor is no longer effective. Sometimes another even smaller capacitor is added. Thus, the main filter capacitor might be a 100 μF aluminum electrolytic type, the second capacitor could be a 10 μF solid-state tantalum type, and the third capacitor could then be a 0.1 μF ceramic type.

- Components C and D are intended to slow the rise and fall of the switching waveform. Although this gets to the heart of noise production, one must be prepared for a sacrifice in the efficiency of the switching regulator. In practice, it is often found that a very small smoothing of the abrupt switching transitions goes a long way in noise reduction. In general, the less reliance upon this method, the better. The improvement in switcher performance attained in recent years has been attributable to faster response in the switching transistor and free-wheeling diode.

- Capacitor E and low-pass filter F serve similar purposes in reducing noise feedback into the unregulated supply. Since much of this noise is actually caused by the switching transistor, capacitor

E acts as an energy reservoir and carries a high ripple current. Its grounding, together with that of the free-wheeling diode, is generally made separate from all other components; with poor grounding, long leads, or bad physical placement, this capacitor can be the source of much trouble, because an inductive loop can be formed to feed noise into both the input and output of the regulator. Such an inductive loop can also be responsible for erratic switching behavior.

- Filter *F* is usually employed to attenuate higher frequencies, particularly when there is an appreciable distance between the unregulated supply and the switching transistor. A feedthrough type is often used. Excessive inductance is dangerous in this circuit position because of the negative-resistance characteristic of the switching transistor; oscillation and other instabilities occur all too easily.

- Resistance *G* tends to make the input current to the rectifier more sinusoidal rather than pulse-like. This measure lowers the efficiency of the rectification system.

- Another source of rectification noise is due to the reverse-recovery characteristic of the rectifying diodes. For 60 Hz, it may prove unwise to use diodes classified as *fast-recovery* types. In more complex switching-type regulators, where fast-recovery diodes must be used to achieve high rectification efficiency, one has a choice of "soft" or "abrupt" recovery types. Where the Schottky diode is applicable, one can obtain virtual freedom from this source of noise.

- Power-line filter elements *H* are employed to prevent switcher noise from entering the power line. This is highly desirable, for many noise problems stem from conduction and radiation of noise through the power line. Such noise is often routed from one part of a system to another, and defies remedies generally successful with more direct paths.

The physical placement and orientation of the switching supply, as well connection to the common system ground and merits investigation. Supplies which incorporate toroidal magnetic components are advantageous when used in close proximity to sensitive circuits. Even ventilation holes and apertures in the power supply's cover may permit leakage of high-frequency noise. (Even the shielding capability of coaxial cable varies widely, depending upon the mesh, thickness, and conductivity of the outer sheath.)

NOISE GENERATION IN THYRISTOR CIRCUITS

The abrupt turnon of thyristors generates harmonics, which have sufficient energy to interfere with many types of systems. This energy extends well into the megahertz range, even when the switching rate is only 60 Hz. Inasmuch as the amplitudes of the higher harmonics approaches zero asymptotically, interference can occur in radio and TV receivers. Turnoff also contributes to the electrical noise generated by phase-controlled thyristor circuits. The noise, however produced, gets into equipment by both conduction and radiation. It is particularly desirable to keep the noise from entering the AC power lines, for it then tends to radiate, finding its way into sensitive circuitry.

Figure 4-3 shows how a low-pass filter may be used in a triac circuit, to prevent the noise both from being passed into the load and from entering the power line. The configuration of the filter is primarily for attenuation of outgoing energy, but

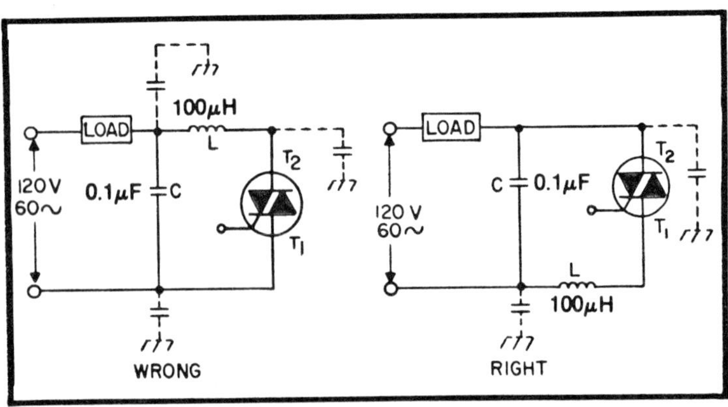

Fig. 4-3. Application of low-pass line filter to a triac circuit. The L and C values shown are typically used in the 100 to 1000 watt range of power levels. The cutoff frequency is approximately 50 kHz.

such a filter also protects the triac from incoming noise. The performance of the filter is degraded by the connection method shown in Fig. 4-3A, because the relatively high capacitance at terminal 2 of the triac tends to bypass noise frequencies around the inductor. In B, however, this capacitance is advantageously used, for it now acts as the shunt arm of a pi filter.

The simple filter configuration of Fig. 4-3B often appears in phase-control circuit utilizing triacs and SCRs. Not infrequently, the noise generation is actually increased. What generally happens is that the filter is shock-excited, and the ringing that results momentarily turns the thyristor off. It quickly turns on again, and the load may not be affected by the momentary power lapse, but noise is generated by such erratic operation. In fact, the higher the Q of the filter, the more likely this may be to happen. Lowering the Q of the inductor can be helpful, but this technique is self-defeating in that it quickly slows down the attenuation of the filter in its rejection band. A better approach is shown in Figs. 4-4 and 4-5, for both triacs and SCRs. An additional RC combination is added to the shunt arm of the filter; this produces the required damping, but is not nearly so detrimental to the performance of the filter as is the reduction in the Q of the inductor.

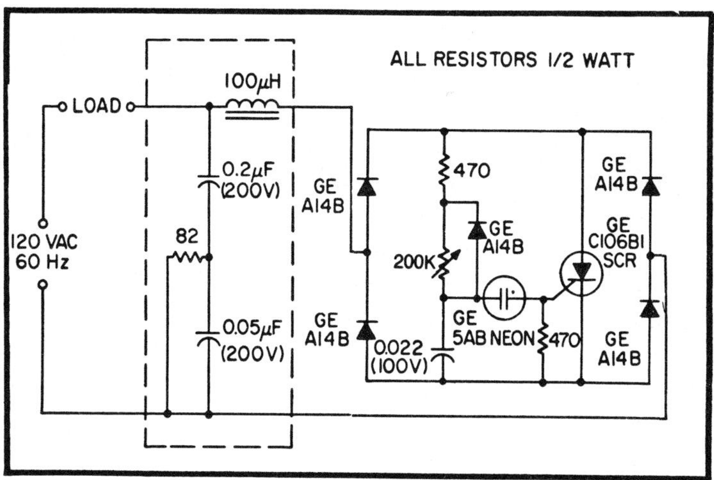

Fig. 4-4. Noise-reduction RFI filter used in full-wave SCR circuit with phase-controlled load power. (Courtesy General Electric Semiconductor Products)

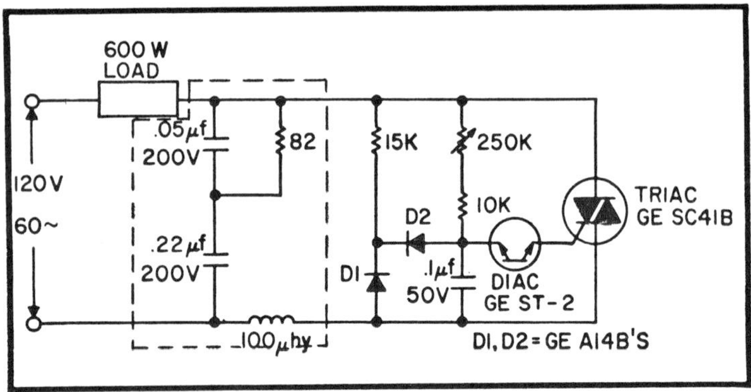

Fig. 4-5. RFI filter can also be used with triac circuits, as in this phase-controlled circuit, equivalent to the SCR version appearing in Fig. 4-4. (Courtesy General Electric Semiconductor Products)

ELECTRICAL NOISE SOURCES

There is more to switching than the mere making and breaking of a switch. So far, we have discussed the topic of noise in terms of its remedies rather than its causes. Since noise originates as a direct consequence of the switching process, it is time that we examined the manner in which noise is produced. While this requires a certain amount of mathematical theory, this aspect of the discussion is kept at a minimum, for much more can be gained from a qualitative approach than from a purely mathematical expression.

First, let us consider the circuit in which the direct current to a resistive load can be made and broken by a single-pole, single-throw switch, as in Fig. 4-6. Suppose this switch can be operated at various rates, and that we shall not now concern ourselves with such secondary effects as arcing, bouncing, or with finite contact resistance. We have then a "perfect" switch. When it is open, it offers infinite resistance to the flow of current; at closure, its ohmic resistance is zero. And just to make the situation ideally simple, let it be assumed that the transition between its two states is accomplished very, very quickly, compared to the *on* or *off* time—even at high switching rates.

If we observed the voltage waveform across the load resistance, we would see a nice, clean square or rectangular wave. But just how "clean" is this train of pulses? The number

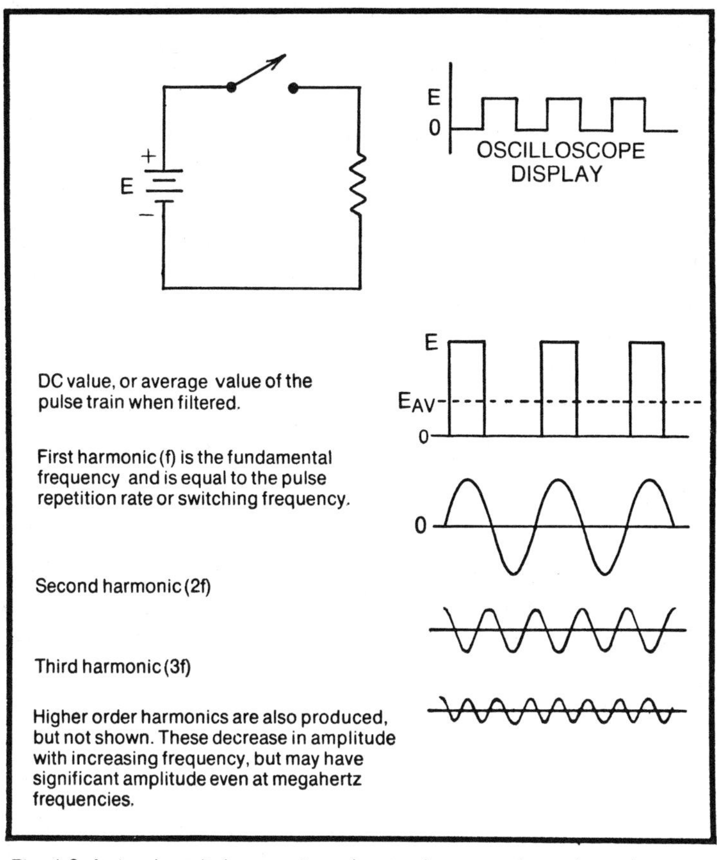

Fig. 4-6. A simple switch cannot produce a clean waveform, for only a pure sine wave (or cosine wave) can be considered to be electronically clean. Other waveforms possess harmonics of the fundamental frequency, and these harmonics are the primary source of noise in electronic circuits.

of pulses per second is exactly that corresponding to the number of times per second we place the switch in its *on* position. And if the switching process is exceedingly well controlled, we could produce a train of uniformly wide pulses. Inasmuch as the oscilloscope screen displays this repetition rate or frequency, and shows no transients or other frequencies, have we not demonstrated a "clean" switching process?

The answer is both yes and no. Yes, the switching operation is as clean as it can be, devoid of transients, spikes, bounce, arcing, wave distortion, etc. And no, there cannot be a nonsinusoidal wave that is mathematically or electronically clean; for according to the important mathematical principle known as the Fourier theorem, all waveforms except the sine wave possess harmonics of the fundamental frequency. All of these harmonics are pure sine waves (or identically shaped cosine waves). Any waveshape can be resolved into a number of sine waves, each harmonic frequency having its unique amplitude and phase relationship. Conversely, we can synthesize any waveshape by appropriately combining such harmonically related sine waves. Rectangular switching waves in power supplies have *many* harmonics; the same is ture of SCR and triac waveshapes.

Figure 4-7 provides extended insight into the consequences of making and breaking the circuit at a periodic rate. This is a universal graph, depicting harmonic production as a function of both pulse rate and pulse duration. We see that the *spacing* of the harmonics is governed by the pulse repetition rate; the

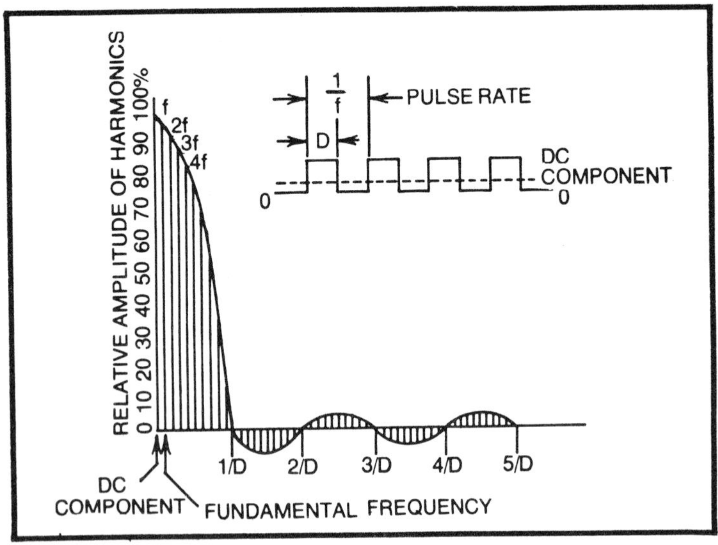

Fig. 4-7. The frequency spectrum of a simple switched waveform. This is a universal curve that can be applied to any pulse repetition rate f and any pulse duration D (rectangular pulses only). It can be seen that the narrower the pulse (or on time of the switch), the greater the number of harmonics and the closer they are spaced. But regardless of pulse width, the zero-amplitude points always occur at frequencies equal to 1/D, 2/D, 3/D, etc. The negative amplitudes are indicative of the relative phase of the harmonics—they are just as "real" as those with positive amplitudes.

actual harmonic *frequencies* are inversely proportional to the pulse duration, or width. Those groups of harmonics which are shown as being negative are simply 180° out of phase with the harmonics contained in the first group. But *all* harmonics have *positive* energy and manifest themselves as signals that radiate, causing interference to susceptible systems.

Note that the successive groups of harmonics grow smaller as the harmonics they contain become higher in frequency, but they never become zero except at periodic zero crossings, such as $3/D$, $4/D$, $5/D$, etc. In practice, substantially zero amplitudes could be achieved through the low-pass action of stray circuit capacitance and resistance. Nevertheless, interference at very high frequencies can be produced by such simple switches, particularly when they are used to interrupt high energy levels or voltages. It should also be pointed out that very nonsimple consequences result when the load is something other than pure ohmic resistance; indeed, the counter electromotive force associated with the interruption of current in an inductive circuit can produce a very substantial amount of noise.

In summary, it can be said that the switching or interruption of currents and voltages in a circuit always give rise to potential sources of objectionable noise. Nevertheless, the desirability of a near dissipationless power supply has stimulated the development of numerous—and often quite simple—techniques of noise reduction. As a result, switching-type power supplies often achieve much lower noise levels than comparable dissipative types using linear control methods. This accomplishment probably results from the fact that designers of switching-type supplies are much more aware of both the causes and cures of electrical noise, and so are not likely to fall into the traps that plague many dissipative designs.

ACOUSTIC NOISE SOURCES

Metnion should also be made of acoustic noise in power supplies. Such noise usually emanates from electromagnetic devices such as inductors and transformers, but can also be produced by capacitors and current-carrying conductors.

All electromagnetic devices tend to emit acoustical noise. In conventional 60 Hz supplies, the audible noise is primarily 120 Hz, or double the line frequency. The source of the noise is derived from the electromechanical stresses exerted within the components, between the current-carrying conductors and the magnetic core materials, at each positive and negative peak of the current. This is particularly prevalent in nickel alloys which tend to exhibit a characteristic referred to as magnetostriction, resulting in periodic elongation and constriction of the magnetic material in response to varying magnetic flux. However, an even greater source of noise in electromagnetic components is due to the shock excitement of the windings and laminations, which tend to ring or vibrate at much higher frequencies than the electrical currents flowing through them. Considering that the human ear is especially sensitive to audio frequencies in the 1−5 kHz range, even 60 Hz electromagnetic devices can produce considerable harmonic noise concentrations (a problem that plagues the fluorescent lighting industry). In switching-type supplies, the acoustic noise problems are often solved by simply raising the operating frequency up to the 20−30 kHz range, above the range of human hearing. (The electrical consequences of this method will be discussed later.) At lower switching frequencies, some benefit is derived from the fact that the magnetic components are much smaller, and the use of toroidal and ferrite cores is much quieter than laminated steel cores. And the lower dissipation of switchers means that shock mounts and standoffs can be used to prevent acoustical coupling, since there is less need for a solid contact to the chassis for thermal heatsinking.

Other sources of noise generation, such as capacitors and conductors, also benefit from the smaller size requirements afforded by switching-type supplies. The use of aluminum chassis and structural parts also decreases the possibility of electromagnetically induced noise, as does the use of sound-deadening foams and potting compounds. As a result, the acoustic noise resulting in a switching-type power supply is no worse than in a dissipative supply—and at high frequencies, may be considerably better.

5
Switching
AC Voltages

Most power supplies are line operated. The output of such supplies may be either AC or DC, but the initial problem encountered is that of how to switch or control the AC voltage. In preceding chapters, we have introduced three basic switching devices that may be used with AC: the diode, the SCR, and the triac. In this chapter, we will examine these three switching elements with their associated circuitry.

THE HALF-WAVE DIODE RECTIFIER

It is inevitable that the half-wave rectifier be considered, since both the circuit configuration and the switching process appear fairly simple. The instrumentation and interpretation of AC and DC readings, however, require an understanding of how waveshapes relate to the concepts of *effective* and *average* values. As shown in Fig. 5-1, the readings of the DC and AC meters are not what the unversed in AC theory would expect, but the measuring instruments, after all, do give reliable readings of the quantities they monitor. When these readings are correctly interpreted, design and servicing become a predictable cause-and-effect relationship. The fact remains, however, that the switching circuit in Fig. 5-1 generates a nonsinusoidal waveform, since segments or

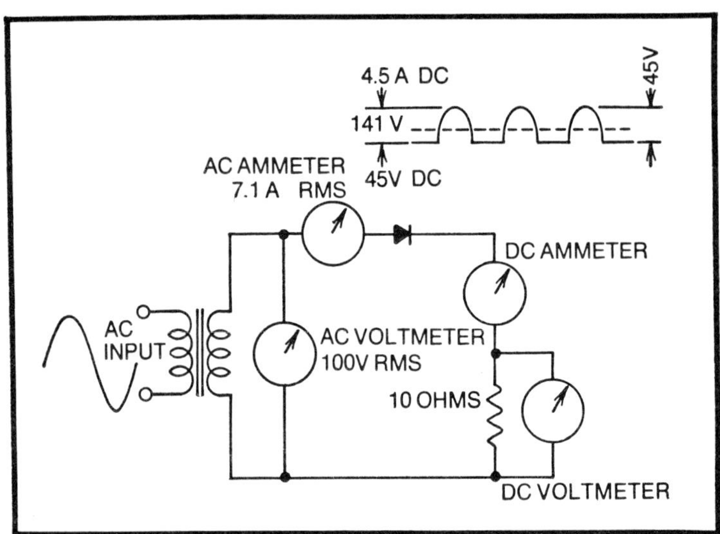

Fig. 5-1. The half-wave rectifier, by itself, is not a very efficient converter of energy from AC to DC. While it is true that the AC input power is equal to the total output power, the output actually consists of a pure DC component and AC harmonics. Thus, in terms of conversion efficiency, the half-wave rectifier is only about 40% efficient, since the AC input power is $7.1^2 \times 10 = 500W$, and the DC component of the output power is $4.5^2 \times 10 = 200W$. In other words, the output of the half-wave rectifier is 40% and 60% AC.

fractions of complete sine waves are not considered to be sine waves at all. Therefore, the half-wave rectifier is a harmonic generator. In practice, harmonic generation in such a circuit can be more severe than would be anticipated by a Fourier analysis of sine pulses. The reason is that the DC component of this wave train can produce varying degrees of saturation in the core of the transformer. Considerable overdesign of the transformer is necessary with respect to rectification if we wish to avoid the effects of operation in the region of core saturation. Unfortunately, overdesign is also required to compensate the low utilization of the transformer when subjected to this waveform, for the harmonics increase the hysteresis and eddy-current losses in the transformer core.

Regulation

An added disadvantage to the simple half-wave rectifier is that it is unable to regulate or control its output voltage; that is, the output voltage delivered to the load varies in relation to the unregulated AC input voltage. Even more complex rectification systems, such as full-wave rectifiers with complicated LC filters, are unable to provide significant regulation of the output voltage. (A possible exception in such systems is the use of voltage-regulating transformers that are able to exercise control over the output voltage through specially designed saturable cores and tuned windings.) Thus, while the half-wave rectifier proves to be a simple AC switch, it also proves to be an impractical switch, inasmuch as there is no way to control its operation to effect voltage regulation.

Conversion Efficiency

The half-wave rectifier is, admittedly, an oversimplification, but it is quite useful in illustrating the relative inefficiency of primitive rectification systems. As Fig. 5-1 indicates, the half-wave rectifier exhibits an AC-to-DC conversion efficiency of about 40%. Full-wave rectifiers, with double the number of rectification pulses, have efficiencies of about 80%, a substantial increase. The addition of capacitive and inductive filtering components can further increase this efficiency. The fact remains, however, that an increase in efficiency is accomplished only with an increase in cost and complexity.

The significance of this efficiency factor is depicted in the waveform of Fig. 5-2, which is of the type obtained from a full-wave rectifier system having a capacitive-input filter for temporary energy storage between rectification pulses. In electronic systems, which are by far the most common, the efficiency of the power supply is affected primarily by the portion of the unregulated input power that is wasted in the regulation process. Thus, in Fig. 5-2, the amount of *usable* power is contained only in the lower rectangular portion of the waveform, which represents the regulated DC output voltage (V_{OUT}) of the power supply. The upper portion of the waveform, representing the AC ripple voltage (V_R) and voltage margin (V_M), is wasted energy that greatly reduces the efficiency of dissipative-type linear power supplies. The ripple voltage can only be reduced by increasing the amount of

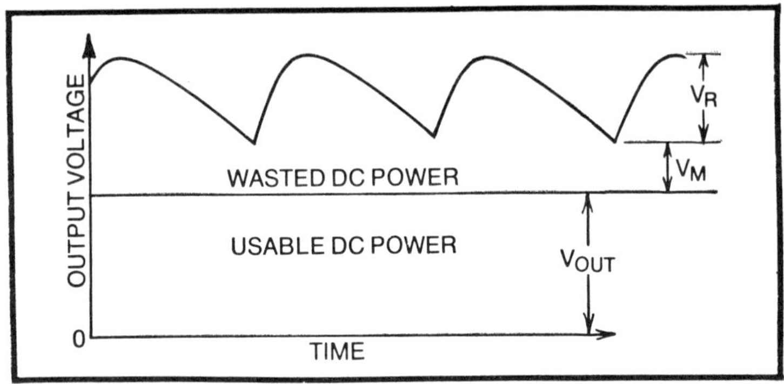

Fig. 5-2. An unregulated input-voltage waveform found in typical dissipative-type voltage regulating systems. The portion of the waveform in excess of the output voltage (V_{OUT}) of the supply represents wasted power that must be dissipated by the series-pass element of the regulator. Modern, switching-type power supplies make use of the entire waveform, and so are significantly more efficient.

filtering, and hence the cost of the power supply. The voltage margin is a function of the variations in AC input voltage and the minimum voltage drop required across the series-pass regulating element, representing a worst-case design constraint.

Switching-type power supplies make use of the entire input waveform depicted in Fig. 5-2, since *all* of the waveform represents "usable" power. A switching regulator, for example, would compensate for the ripple-voltage component of the waveform—not by dissipating it, but by varying the pulse rate or pulse duration of its switching process to convert the ripple voltage into usable output power. Similarly, the voltage margin, which varies with fluctuations in AC line voltage, presents no problem to the switching regulator, for the switching regulator also varies its pulse rate or pulse duration to compensate for such fluctuations. As a result, switching-type power supplies generally place lesser demands upon their input rectification systems than would be the case with dissipative supplies, and this tends to reduce the size of the power supply and compensate for the cost of the switching components.

THE SCR SWITCH

The SCR switch resembles the half-wave rectifier, but has the advantage that it is controllable. The circuit, shown in Fig.

5-3, also contains instrumentation. This switch is not only a generator of prolific harmonics but also causes the power factor of its AC feed line to be low, despite the fact that the load may be pure resistance. (The SCR circuit behaves as an inductive reactance.) In Fig. 5-3, the power factor becomes progressively worse (lower) as the conduction angle is made smaller.

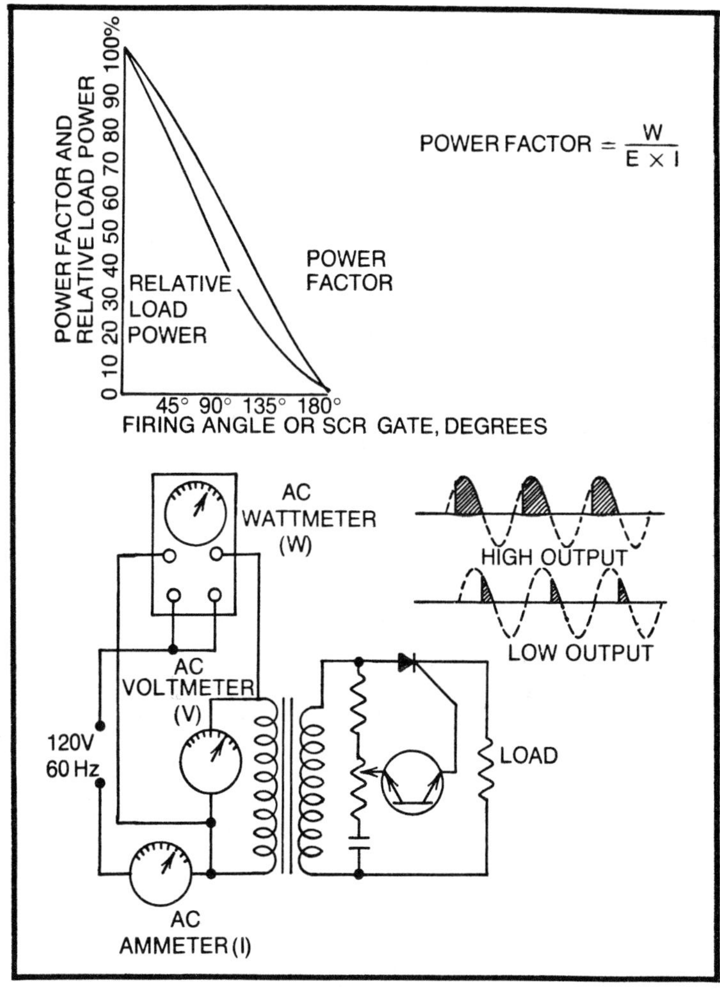

Fig. 5-3. The SCR switching circuit is able to control or regulate the amount of power delivered to the load, but the method of controlling the power by varying the phase angle also decreases the power factor of the supply and greatly contributes to the generation of electrical noise.

Incidentally, the use of a conventional power-factor meter provides yet another example of complexity arising from apparent simplicity. It so happens that this instrument responds to the phase difference between the fundamental voltage and current. But when appreciable harmonic energy is present, as with SCR switching, an erroneous reading is obtained—the departure from accuracy can exceed 100%.

When the SCR switch is delivering low output (corresponding to a small angle of conduction), it becomes somewhat nebulous to refer to the switching process as one of high efficiency. It may be true that relatively little dissipative power is involved in either the *on* or *off* positions of the switch, but the high current consumed through the power line because of low power factor can itself be the mechanism through which appreciable power is lost.

The phase-controlled switching method employed in such circuits creates very abrupt transitions in the output waveform, and this greatly increases the magnitude of noise-producing harmonics. The DC conversion efficiency of

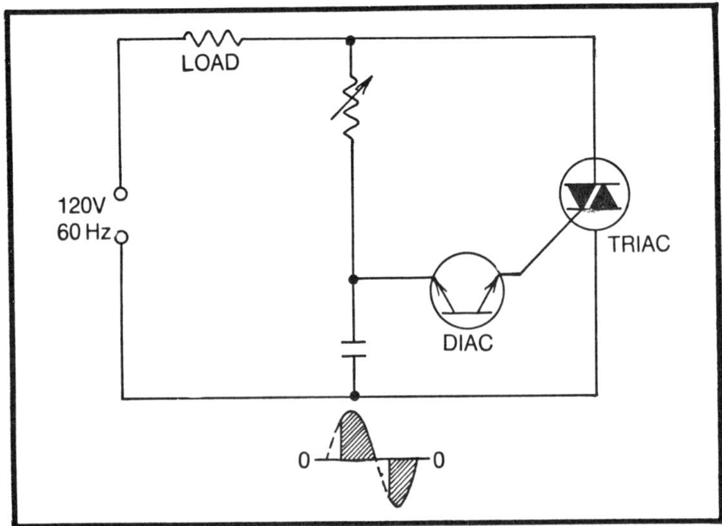

Fig. 5-4. Full-wave, phase-controlled triac. The symmetrical switching produced by this arrangement has considerable appeal, but noise generation is a problem that still must be coped with. Nevertheless, such switching circuits (found in most light dimmers) prove to be a simple and inexpensive way to regulate power to AC loads. Such circuits are not very practical in obtaining DC power for electronic circuits.

the SCR system is no better than that of the half-wave rectifier; if anything, it is worse. Thus, while the SCR circuit makes a good AC switch, inasmuch as it is able to control the output power level, it is a bad switch with regard to power factor, conversion efficiency, and noise generation.

THE TRIAC SWITCH

The triac switching process is a push-pull or full-wave version of the SCR switching mode (Fig. 5-4). Its propensity for generating noise is similar to that of the SCR, for its harmonic spectrum is also extensive. If the two sections of the triac are balanced with respect to firing potential, there will be a tendency towards cancellation of the DC component flowing through the supply transformer. However, the low power factor which accompanies small angles of conduction is not overcome by use of this thyristor.

THE ZERO-CROSSING SWITCH

The block of Fig. 5-5 constitutes a switch which, by virtue of its switching process, produces few operational side effects. Such a switch makes and breaks the circuit without generating significant harmonics, altering power factor, or saturating the supply transformer (if one is used). This switch does not produce an erratically varying inrush current, as ordinarily results when a transformer is suddenly energized from an AC source. At the same time, this switching process utilizes the dissipationless principle of switch control to full advantage.

The circuitry illustrated in Fig. 5-5 is used to switch the triac on and off at the precise instant that the AC input waveform crosses through its zero-voltage point. The advantage of this approach is derived from the Fourier postulate that the sine wave is the major building block of all waveshapes. By switching at the zero crossing of the input sine wave, abrupt switching transitions are prevented; only the sine wave passes through to the load. (In truth, a small transition does occur in going from the *off* state to the initial ramp of the sine wave, but the noise produced at this instant is negligible when compared to the sharp transition encountered when attempting to switch the waveform at any other time during the cycle.)

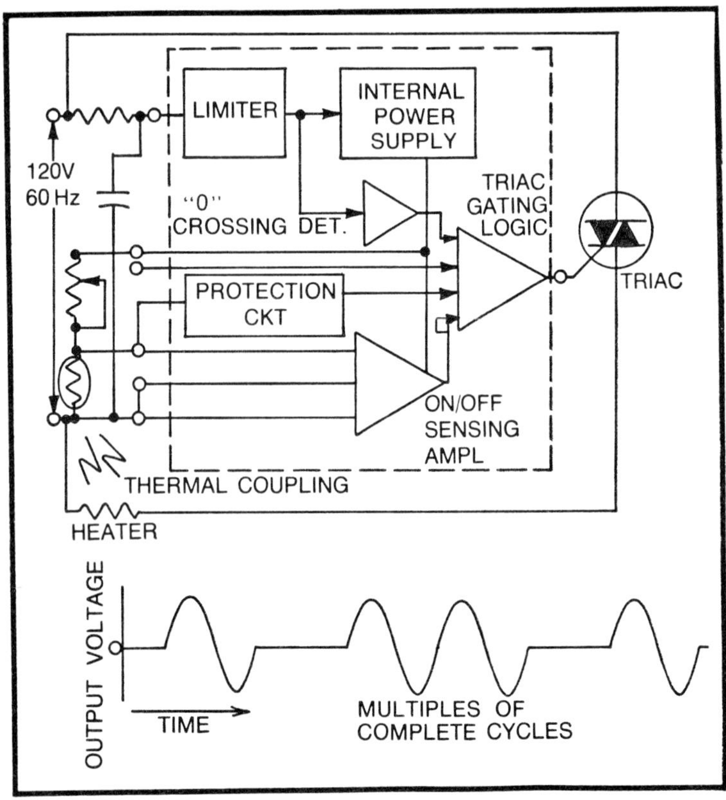

Fig. 5-5. A zero-crossing switch is a very effective circuit for controlling AC loads that do not require fast response times. Circuit operation is limited to delivering one or more complete sine waves to the load, a process that practically eliminates noise generation but does not permit fast response to varying loads. The circuitry within the dashed lines is contained within an integrated circuit.

The arrangement illustrated in Fig. 5-5 actually comprises a switching regulator—although not the type generally used for powering electronic equipment, since it does not convert the AC input voltage into a DC output. Still, this type proves to be extremely useful in regulating many AC loads, such as heating elements in temperature-controlled ovens. In this latter example, a thermistor might be used as the temperature sensor for activating the triac circuit. Temperature regulation would occur through the action of the triac circuit in delivering multiples of complete sine waves to the heating element in the oven. The response time of such a regulator is relatively slow, but the requirements for this and similar applications rarely require high-speed regulation. The advantages of such a system are high efficiency, relative simplicity, and low noise generation which permits close proximity to sensitive electronic equipment.

Although the zero-crossing mode of switching, when applied to sine waves, proves to be the "cleanest" technique, it does not enjoy widespread use in actual switching regulators. Such switching slows down the response time, which already suffers poorly in comparison with that of linear regulators. If a 20 kHz sine wave is applied to a load in groups of complete cycles averaging, say, 10 cycles per group, it is as though we were dealing with a 2 kHz wave insofar as response time is concerned—the response time *cannot* be faster than the duration of one group of cycles. While a much higher-frequency version of this type may be implemented in the future, at present such slow response is entirely inadequate for modern DC-powered electronic equipment.

6

Switching
DC Voltages

One of the most favored techniques for switching DC voltages involves an electronic equivalent of the single-pole/single-throw mechanical switch. Interruption of the current flowing through this type of device is responsible for much noise generation, but such effects can be suppressed with a wide variety of relatively simple filters. The switching circuit shown in Fig. 6-1A produces a pulsed voltage across the load, and though it is possible to define an *average value* for the pulse train, it is no more than a concept. Except for a few nondemanding applications, such as charging storage batteries, the equivalence between the pulsed power and the average level is more academic than practical.

By means of a capacitor, a considerable improvement is brought about in Fig. 6-1B. The sawtooth wave comprises segments of exponential charges and discharges of the filter capacitor. If the capacitor is large enough, the minimum value of the wave never dips to zero, and a truly direct current passes through the load. This simple scheme has much to recommend it, but it leaves much to be desired. For example, if we wish to diminish the AC ripple voltage superimposed upon the DC level, an impractically oversized capacitor is required. Aside from the expense and massive proportions, such a large capacitor would have an inordinate slowdown effect on the

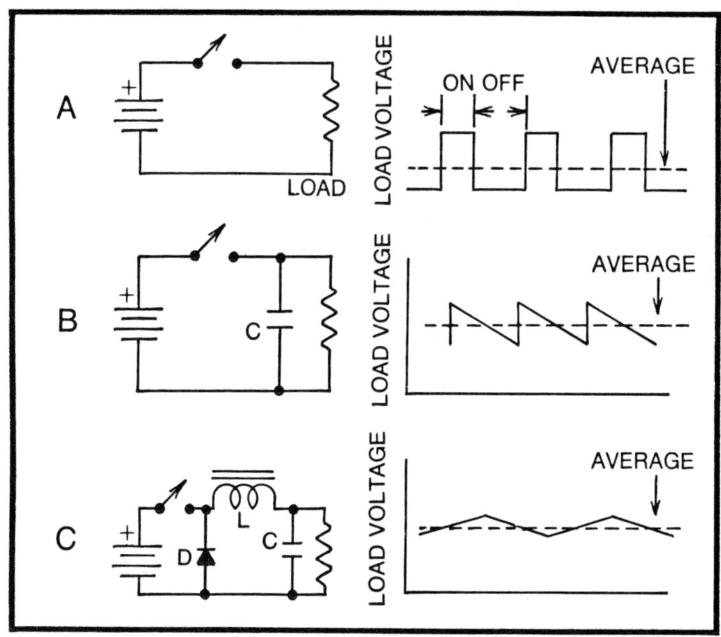

Fig. 6-1. Evolution of DC switching circuits. (A) The waveform of this switching circuit is too intermittent: the "average" is a conceptual value and is not experienced as such by the load. (B) A fairly sustained voltage is applied to the load using a capacitive filter. (C) This is a very usable arrangement; the diode provides for the near-continuous flow of load current throughout the switching cycle, and at the same time, inductive arcing is suppressed when the switch is opened.

response time of the regulator. Of course, the switching rate could be speeded up, also causing the ripple amplitude to decrease, but only so much enhancement of filtering performance can be secured in this manner, for ultimately the dissipative losses in the switch itself begin to mount up. Instantaneous currents into the capacitor tend to be high, and these are similar to the surge current in rectifiers that flow when a power supply is first turned on.

LC FILTERS AND FREE-WHEELING DIODES

It is only natural to think of adding an inductor. Unfortunately, an inductor by itself is useless for chopped currents, since the opening of the switch must release the energy previously stored in the magnetic field of the inductor, which causes arcing and other manifestations of high voltage.

A mechanical switch might have its contacts burned or welded together, and it is needless to detail the damage likely to overtake a semiconductor switching element.

Suppose that, for the purpose of arc suppression, a diode is connected as shown in Fig. 6-1C. The diode does not interfere with the injection of pulses into the inductor, and is so polarized as to absorb the high-voltage transient generated when the circuit is interrupted. Although such applications of diodes are not new, in this particular circuit it turns out that the diode also helps bring about a steady delivery of power to the load. When the switch is open, the load current is then supplied by the stored energy in the magnetic field of the inductor, with the diode making a complete path for the current to flow.

A more relevant way to view the diode is a means whereby the energy in the magnetic field of the inductor (stored during the previous *on* time of the switch) is converted into useful current flow in the load. Without synchronizing techniques, or any additional circuitry, the diode provides for this current flow at the right time—when the switch is in its *off* position. The diode enables the LC filter to perform in an acceptable manner; and because of the diode, we should become accustomed to thinking of the LC filter as an *energy reservoir* rather than merely a frequency-selective network. Other names for this diode are *catcher* diode, *coasting* diode, *bypass* diode, *flywheel* diode, and *commutating* diode.

SWITCHING AND VOLTAGE REGULATION

With a few simple components—a switch, an inductor, a capacitor, and a diode—the electrical power available from a DC source can be interrupted, smoothed again, and then applied to a load in the form of essentially pure DC power. Thus far, we have assumed that the duty cycle and frequency of the switching process were nonvarying. We know from Chapter 2 that this is not so in actual regulating supplies, for the effect of varying the pulse width and repitition rate is what permits us to control the output voltage of the regulator. This is illustrated in Fig. 6-2. When such control is made automatic by means of an appropriate sensing technique and feedback loop, the result is regulation of the DC voltage applied to the load.

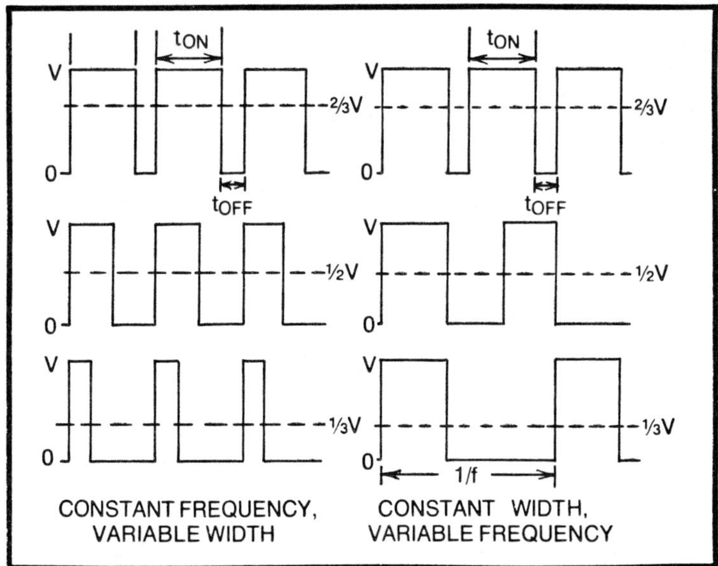

Fig. 6-2. Control of the output voltage can be obtained by either of two basic methods, as illustrated above. Some regulators incorporate a combination of both methods in order to place minimum and maximum limits on operating frequencies and pulse widths.

The two methods of control have much in common. The output voltage decreases in situations when the *ratio* of *on* time to *off* time is decreased; this can be brought about by decreasing either frequency or pulse width (*on* time). Thus, the two sets of switching waveforms in Fig. 6-2 represent equivalent methods of bringing about a decrease in output voltage, even though different approaches are used. This can be described by the equation:

$$V_{OUT} = \left(\frac{t_{ON}}{t_{ON} + t_{OFF}} \right) V_{IN}$$

Note that the denominator of this equation represents the time for one full switching cycle. If the switch remained closed, the output voltage would equal the input voltage; if the switch remained open, the output voltage would be zero. Obviously, actual output voltages will be somewhere between these two extremes. It is not easy to say that better results should be forthcoming from either the constant-width or the constant-frequency technique; excellent performance has been abtained with both methods, and much depends upon the skill with which the designer implements the control function.

In self-excited switching regulators, both frequency and *on* time are allowed to vary. Which parameter is more influential in bringing about regulation depends upon many factors. Although such random variation of the switching waveform leads to entirely satisfactory regulation, many designers feel it is better engineering practice to operate with either the frequency or the *on* time constant. Some restrictions on these parameters are justified, from a design standpoint, in order to place safe limits on the operation of the regulator. In many circuits, if the *on* time is too short or the frequency too high, the series-pass transistor begins to dissipate excessive energy; and if the *on* time is too long, excessive currents may damage the transistor or saturate the inductor.

The possibility of using a constant *off* switching mode has not, thus far, been mentioned. This mode can be viewed as a degradation of the constant-frequency mode. In both modes, the variable parameter is the pulse width, or *on* time. In the constant-frequency mode, the denominator of the fraction is held constant, whereas in the constant *off* mode, that part of the fraction is allowed to change as *on* time is varied. The net result is that with variations in *on* time, the value of the whole fraction does not change as fast in the constant *off* mode as it does in the constant *on* mode. In order to produce maximum regulating effect, the value of the fraction should be sensitive to changes occurring in the variable parameter, and the constant *off* mode is definitely not as sensitive. But a boost for the constant *off* mode involves the fact that digital waveforms from logic circuits can sometimes be conveniently implemented in this fashion.

Other things being equal, the higher the switching rate the better, since we wind up with smaller inductors, capacitors, and transformers. However, other things do not remain equal—switching losses go up with faster switching rates because of the greater time consumed in achieving turnon and turnoff, electrolytic capacitors become less effective, and losses tend to mount in the free-wheeling diode. In many designs, frequencies in the 20–30 kHz region are favored

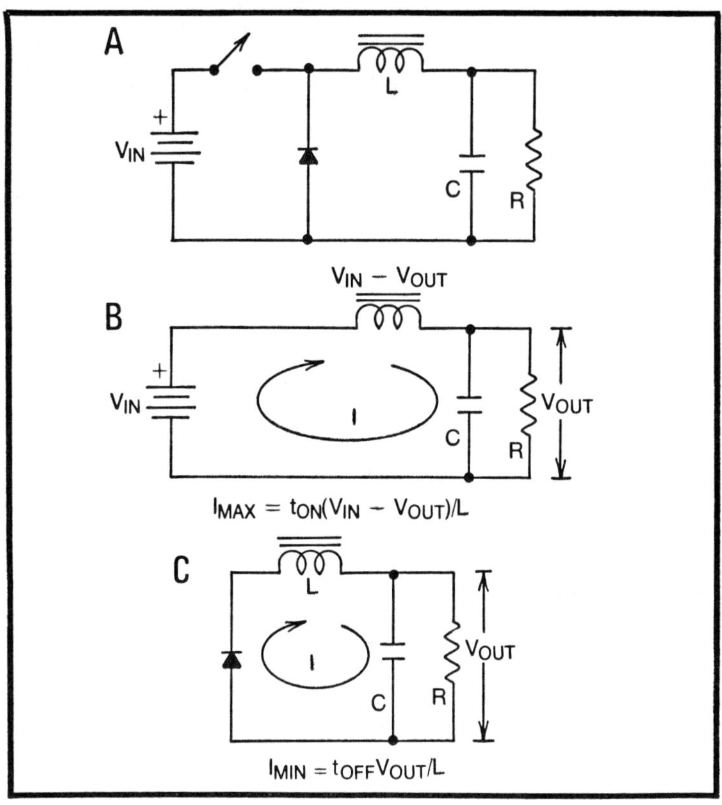

Fig. 6-3. The current peaks in the inductor are functions of on and off times. (A) The basic switching circuit. (B) Current path during ON time. (C) Current path during OFF time. The equations shown are valid when current is expressed in amperes, time in seconds, and inductance in henrys.

because this region is well above audible frequencies, but still represent an acceptable tradeoff between efficiency and economics. In the future, switching rates of 50 kHz and greater will see much more use. On the other hand, where audible noise does not tend to be objectionable, reasonable overall performance is often obtained at several kilohertz.

INDUCTOR CURRENT

A key to the operation of the basic switching circuit shown in Fig. 6-3 is what happens to the current passing through the inductor. Because the inductor is subjected to step-function voltages, it is easy to determine the current through it. For small changes in V_{OUT}, the current through the inductor is just a waveform made up of ramps or linear slopes. The peak amplitudes of such a wave can be calculated from purely geometric considerations. Referring to both Figs. 6-3 and 6-4, we see that

$$I_{MAX} = t_{ON}(V_{IN} - V_{OUT})/L$$

In so many words, the factor $(V_{IN} - V_{OUT})/L$ is the slope of the *on* ramp and, when multiplied by the elapsed *on* time (t_{ON}), produces the highest current (I_{MAX}). A similar procedure is used for the calculation of I_{MIN}. In this case, the *off* slope is represented by the factor E/L.

The important thing about the current through the inductor is that it is essentially of a *sustained* nature, despite its sawtooth waveform. And unlike the inductor current in a simple half-wave rectifier circuit, there is no dip to zero. In the switching circuit, the inductor and the free-wheeling diode work together to produce a near constant current through the load. In Fig. 6-3C, direction of current flow prevailing after the switch turns off is the *same* as during the time the switch remains turned on, even though the voltage across the inductor has reversed.

A ripple voltage is not always in evidence when the output of a switcher is observed, since the electrolytic filter capacitors have appreciable series resistance. The waveform is therefore a composite of the voltages developed across internal resistances and capacitances. For the moment, however, we will deal only with idealized components to establish basic operating mechanisms, although the major challenge in the development of switching-type regulators has been in the manufacture of ideal components. The theory of the switching-type supply has been around for a long time, but the continuing refinement of these supplies is primarily contingent upon still better switching transistors, filter capacitors, free-wheeling diodes, and other critical components.

SWITCH OR TRANSFORMER?

An important aspect of previous discussions pertaining to the basic DC switching circuit is the matter of control.

Specifically, the DC output voltage is governed by the duty cycle of the switching wave. This is true whether the switching speed or the *on* or *off* times of the switching cycle is varied. It is both interesting and instructive to look at this control function.

The output voltage is determined only by the switching parameters and the input voltage. Specifically, we have the important relationship:

$$V_{OUT} = \left(\frac{t_{ON}}{t_{ON} + t_{OFF}} \right) \times V_{IN}$$

where t_{ON} is *on* time, and t_{OFF} is *off* time of the switch. Inasmuch as $t_{ON} + t_{OFF}$ denotes the period of one switching cycle, and period is the reciprocal of frequency or switching rate f, the equation can also be written:

$$V_{OUT} = f t_{ON} V_{IN}$$

To further simplify the form of this basic relationship, we can let the factor $f t_{ON}$ be represented by the letter k. Then the expression for output voltage becomes simply, $V_{OUT} = k V_{IN}$. It is also true that input current I_{IN} and output current I_{OUT} are related by k; that is, $I_{IN} = k I_{OUT}$. I_{IN} is the current that would be measured by inserting a DC ammeter in one of the battery leads of the circuit in Fig. 6-1; and I_{OUT} is the same quantity as previously designated I_{L}, the inductor current, also equal to the load current.

Although we have dealt with the current in the inductor, this is the first mention of the currents flowing into and out of the switching circuit. It may appear somewhat unnatural that these two currents are *not* the same, but this probably stems from our tendency to think of the switching transistor in much the same fashion as a series-pass transistor in a linear, dissipative regulator. The series-pass transistor in the linear regulator is a "rheostat" and the DC currents flowing into it from the DC source out of it into the load are almost identical. Not so with the switching circuit!

Let us rewrite the voltage and current expressions in terms of their common factor, k; we obtain $k = V_{ON}/V_{IN}$ and $k = I_{IN}/I_{OUT}$. Inasmuch as quantities equal to a common quantity are equal to each other, we can write,

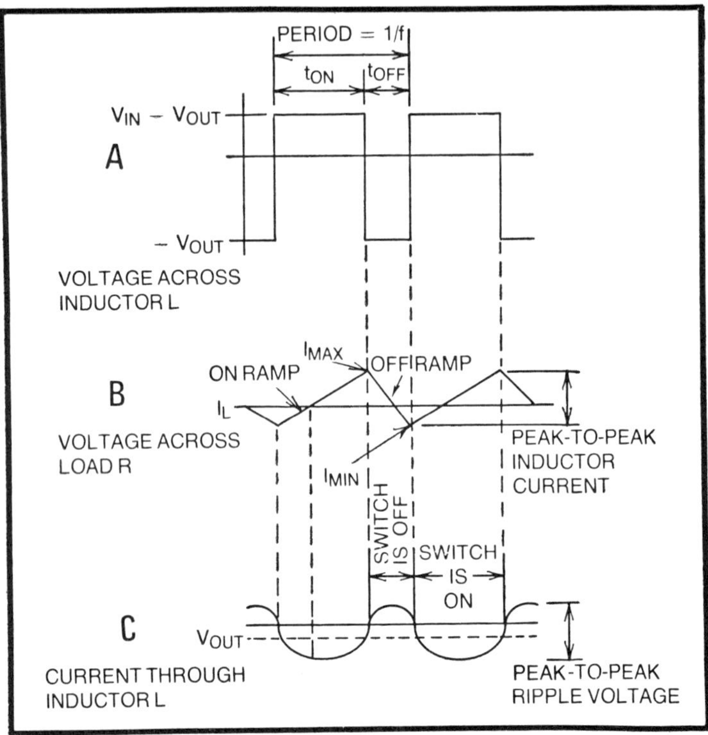

Fig. 6-4. The waveforms associated with the switching circuit of Fig. 6-3. (A) Note that the average voltage across the inductor is zero. (B) The average of the ramp-produced current through the inductor is equal to the load current. (C) This ripple is actually caused by the operation of the switch. Note its phase relationship to the current ramp through the inductor by comparing current and voltage peaks.

$V_{OUT}/V_{IN} = I_{IN}/I_{OUT}$, which can finally be written,

$$V_{IN} = V_{OUT}I_{OUT} \quad \text{or} \quad P_{IN} = P_{OUT}$$

which says that the input power and output power of the switching regulator are identical—and if there were no ohmic losses in the regulator components, the switching-type power supply would have an efficiency of 100%, the same *theoretical* efficiency as a transformer!

In Fig. 6-5, the above-discussed relationships are summarized and compared to the input/output relationships in a simple AC transformer circuit. Here, one sees the justification of describing the basic DC switcher as a "DC

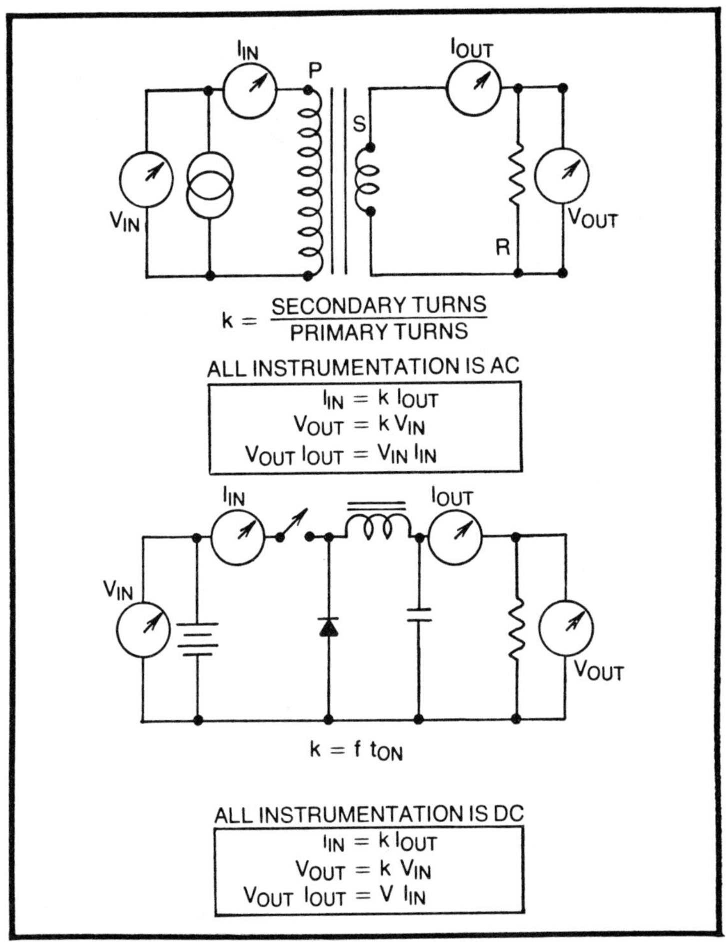

$$k = \frac{\text{SECONDARY TURNS}}{\text{PRIMARY TURNS}}$$

ALL INSTRUMENTATION IS AC

$$I_{IN} = k\, I_{OUT}$$
$$V_{OUT} = k\, V_{IN}$$
$$V_{OUT}\, I_{OUT} = V_{IN}\, I_{IN}$$

$$k = f\, t_{ON}$$

ALL INSTRUMENTATION IS DC

$$I_{IN} = k\, I_{OUT}$$
$$V_{OUT} = k\, V_{IN}$$
$$V_{OUT}\, I_{OUT} = V\, I_{IN}$$

Fig. 6-5. Comparison of AC and DC transformers. The basic switching circuit is a true DC transformer. Although the transformation ratio is derived differently for the AC and DC circuits, the equations in the boxes prove that the same type of relationship connects input and output quantities in both circuits.

transformer." The fundamental feature of ideal transformer action is the ability to change either voltage or current with *no loss of power*. Other methods of changing DC voltage or current are accompanied by power loss. Typical examples are the use of the rheostat or analogous devices such as the series-pass transistor in linear regulators, the potentiometer, and level-changing sources such as the zener diode. In all such methods, a power loss is absorbed by the inserted element, and in none of these examples can *more* current be extracted than that which enters the circuit. Unlike these conventional ways of changing DC voltage level, the basic DC switching circuit has, like the AC transformer, the property of delivering a *greater* current to the load than the current initially injected into the circuit. This is the most important feature of the basic DC switching circuit, and it is directly responsible for the important advantages that switching regulators have over dissipative types.

While the switch undoubtedly makes it possible to construct a DC transformer, the switch itself is not solely responsible, but rather works in conjunction with the energy-storing inductor. The role of the inductor in the circuit is analogous to that of a variac or variable-voltage auto-transformer, except that the inductor is not tapped and cannot, by itself, vary the output voltage. Instead, the switch is the controlling element that determines both how much energy is to be stored in the magnetic field of the inductor and how that energy it to be released to obtain the desired output voltage. Of course, the switch requires additional circuitry to sense the output voltage and control its operation, but the analogy is essentially correct. And, similar to the autotransformer, switching circuits exist which are able to both step up and step down the input voltage, so that output voltages may be either higher or lower than the input.

7

Know Your
Switching Components

A delightful aspect of designing, constructing, or servicing simple unregulated power supplies is the freedom we have to use a wide variety of parts and components. The flexibility extends to the physical layout, and virtually nothing in these primitive supplies is critical once the basic requirements dictated by voltage breakdown and power-handling ratings have been complied with. Often, tradeoffs can be made without incurring significant degradation of performance; for example, if a certain inductance is not available for the filter choke, a considerably smaller or larger one might suffice with only a compensatory increase or decrease in the filter capacitance.

Then came the modern, linear voltage regulator. Because of the more stringent needs of complex circuits and systems, the linear regulator, with its high-gain feedback loops and much tighter specifications on performance parameters, required greater care in the choice of components, in the layout of parts, and consideration of new factors in the operating environment. Deviation from recommended components quickly yielded such plagues as oscillation, latchup, and poor regulation. So, the linear regulator required more experience, more know-how, and more insight into application requirements than did its predecessor, the simple, unregulated power supply.

When the modern, switching-type regulating power supplies came into being, they were accompanied by a great many superior components that took some of the magic out of their design. As a result, even though switching-type supplies are generally more complicated than linear supplies, the use of integrated circuits and special switching components has made switching-type supplies much easier to work with. But to take advantage of these new components, the essential characteristics of these devices must be thoroughly understood.

THE OUTPUT FILTER CAPACITOR

A good place to start is with the output filter capacitor. In the first place, the basic design equations of the switching-type supply clearly reveal that you cannot proceed on the premise, "if a little is good, more is better." Switching frequency, duty

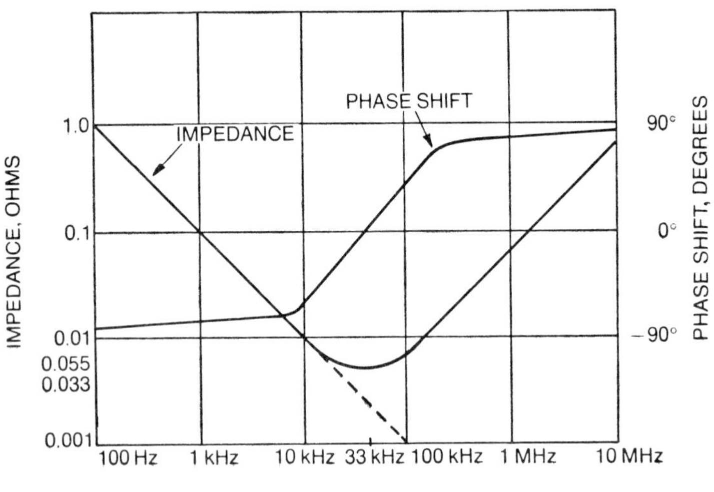

Fig. 7-1. Impedance and phase characteristics of a 1500 μF electrolytic capacitor. The initial straight portion of the impedance plot represents pure capacitive reactance, and the extended projection (dashed line) represents the characteristics of an ideal capacitor, one without either resistance or inductance. The upswing of the curve above 33 kHz results from internal inductive reactance within the capacitor, and this parameter is of prime importance in the design of switching regulators.

cycle, choke inductance, and input and output voltages are all interrelated, so any brute-force approach must be ruled out. It is also important to realize that we get more than we bargained for in electrolytic filter capacitors. That such capacitors also contain resistance and inductance should not come as a surprise, because stray parameters are encountered all the time in electronic designs and parts, and they are frequently ignored. But the stray inductance and internal resistance of electrolytic capacitors are of prime importance in the operation of switching-type power supplies, and therefore cannot be ignored.

Figure 7-1 depicts the impedance of an aluminum electrolytic capacitor as a function of frequency. Note that the plot is made on log/log paper to linearize the reciprocal relationship between frequency and capacitive reactance. This relationship is described by,

$$X_C = \frac{1}{2\pi f C}$$

where

X_C = the capacitive reactance in ohms.
f = the sine-wave frequency in hertz.
C = the capacitance in farads.

The capacitor obeys the capacitive-reactance equation up to about 10 kHz, for up to that frequency the plot is a straight line. The slope, incidentally, is the same for all ideal capacitors and for actual capacitors at low frequencies. Figure 7-1 shows that this slope decreases by a factor of ten for each tenfold increase in frequency, which can also be expressed by stating that the capacitive reactance decreases 6 dB per octave of frequency (20 dB per decade of frequency).

Equivalent Circuit of the Capacitor

We could interpret the circuit characteristic depicted in Fig. 7-1 as a low-Q resonance produced by a series arrangement of capacitance, inductance, and resistance. (See also Fig. 7-2.) Inasmuch as these curves were plotted from measurements obtained with an electrolytic capacitor, we may conclude that, in addition to capacitance, the capacitor

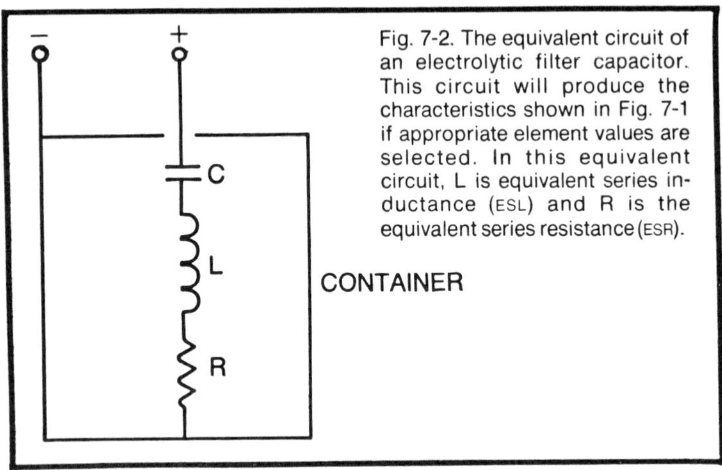

Fig. 7-2. The equivalent circuit of an electrolytic filter capacitor. This circuit will produce the characteristics shown in Fig. 7-1 if appropriate element values are selected. In this equivalent circuit, L is equivalent series inductance (ESL) and R is the equivalent series resistance (ESR).

does contain inductance and resistance. And as the frequency is raised, the inductive reactance of the capacitor's foil and connecting wires eventually dominates. Therefore, the series circuit of Fig. 7-2 is a very down-to-earth representation of the component we refer to as an electrolytic filter capacitor. Most other components can be similarly represented; for example, a grid-dip meter will demonstrate that small ceramic capacitors with their leads shorted together behave as resonant LC circuits, with the inductance in the leads and the sharpness of resonance governed by internal resistance.

The equivalent series R and L are responsible for the following undesirable effects in switching-type power supplies:

- Both the shape and the magnitude of the ripple voltage deviate from that which would be obtained from pure capacitance.
- The design of the switcher is more difficult than would otherwise be the case, since the output ripple of these supplies is one of the design parameters in optimizing the switching frequency, the duty cycle range, and the inductor.
- The phase/gain margin that governs the stability of the feedback loop can turn out to be inadequate because of the uncapacitor-like phase behavior at higher frequencies.

- Although much is made of the DC regulating ability of power supplies, a generally more compelling reason to use regulated power sources is to present a low AC impedance to the load over a wide frequency range. The presence of equivalent series R and L tends to defeat this objective.
- The R value is not easy to pin down, and its variation can be considerable with respect to different brands and aging effects. (This also makes it difficult to manufacture capacitors with compliance to tight specifications.)
- Much of the cost disadvantages attributed to the switching-type supply stem from the need for capacitors with very low R and L values.
- The equivalent series R is quite temperature-dependent, further complicating design and application.

The best expedient is to order special capacitor types made specifically for switching-type power supplies, and only from reputable vendors. Various manufacturing techniques and processes are employed to reduce both R and L; for example, there are "stacked" units having a rectangular packaging configuration and four (rather than two) leads. There are also proprietary fabrication methods which are never detailed in manufacturers' literature, and all such methods usually lead to higher costs. Makers of switching-type supplies have, of course, tried to circumvent the higher-cost specialty capacitors, and one widely used technique is to parallel several smaller capacitors instead of using a single large unit, the results of which are shown in Fig. 7-3.

Interpreting Impedance and Phase

In Fig. 7-1 a dip occurs in the impedance at approximately 33 kHz, at which point the phase shift is zero. This is no accident, for in every series-resonant circuit there is a resonant frequency defined by this condition. The impedance at resonance is called the equivalent series resistance (ESR) of the electrolytic capacitor. For the capacitor of Fig. 7-1, the ESR is about 0.055 ohm. As a rule, the lower the ESR, the better

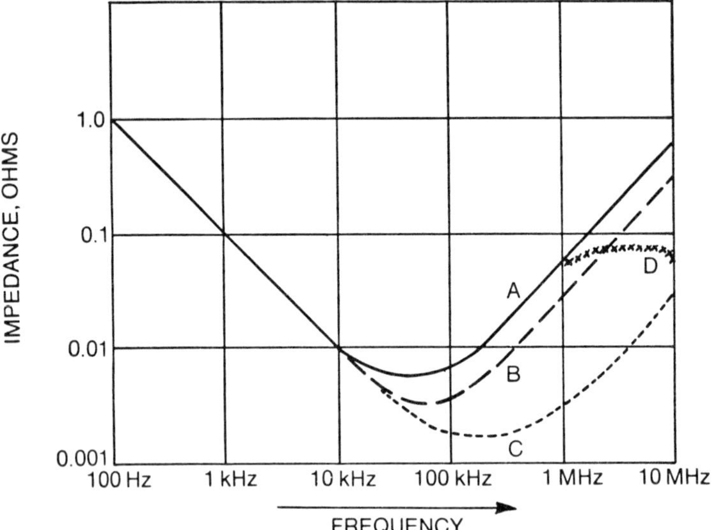

Fig. 7-3. Remedial measures for lowering ESR and ESL in output filter capacitors. (A) Impedance plot of nominally 1500 μF aluminum electrolytic capacitor. (B) Improvement from paralleling several equal-value capacitors to total 1500 μF. (C) Improvement due to the use of a four-lead "stacked" construction capacitor. (D) General nature of improvement in ESL when a relatively small tantalum capacitor is shunted across the 1500 μF unit. For the purpose of decreasing ESL at frequencies greater than several megahertz, ceramic and Mylar capacitors are also used.

suited the capacitor is for the output filter of a switching-type power supply.

The capacitor's impedance is essentially capacitive reactance up to frequencies of about 10 kHz. Then, resistance becomes increasingly predominant, manifesting itself in pure form at the resonant frequency, in the vicinity of 33 kHz. Thereafter, inductive reactance of the leads and capacitor structure begins to dominate at frequencies above 300 kHz.

Although such a way of evaluating electrolytic capacitors is not new, it did not have as much significance for the linear-regulating supply as it does for the switcher. Not only are the characteristics depicted in Fig. 7-1 typical, but the region where the inflection of the impedance curve occurs—in the 10 kHz to 100 kHz range—is of prime importance for switching supplies. Suppose, for example, that this capacitor is used as an output filter for a switcher operating at 20 kHz. For the numerous harmonics of 20 kHz, which are generated by the switching process, the effectiveness of this capacitor is greatly diminished, as its ability to attenuate these harmonics depends upon its ability to bypass or short them to ground. This situation is particularly unfortunate, because the switcher tends to be a prolific generator of high frequency harmonics, and these noisy harmonics in the output not only plague external equipment, but also affect the internal feedback loop used to regulate the output voltage.

The ESR Rating

Because the equivalent series resistance (ESR) exerts a major influence on the impedance of the capacitor in the 10–100 kHz range—the region of optimum efficiency and performance for most switchers—this parameter should be determined as soon as it is known that the voltage rating and capacitance value are suitable. It is becoming increasingly common for manufacturers to supply this information in their technical literature, but it is not, however, always readily available. A much used method of estimating the ESR is by means of the equation: $ESR = P_D/2\pi fC$ where P_D is the dissipation factor, f is the measurement frequency in hertz, and C is the capacitance in farads. The dissipation factor has long been used as a figure or merit for capacitors, and was typically measured at 60 or 120 Hz for filter capacitors, or at 1000 Hz for coupling capacitors. (Note that the measurement frequency f must be known before you can calculate the ESR.) While the D is useful in estimating the ESR of the capacitor, it is useless in estimating the ESL, which may be the dominating factor at switching frequencies.

The ESR actually decreases as temperature is increased. This temperature dependency is particularly strong in the below-zero temperature range, as is shown in Fig. 7-4A. This, together with decreasing capacitance (Fig. 7-4B), has upset the performance of many a switcher that performed well in the environment of the laboratory, but was later expected to work in more rugged environs.

The Ripple-Current Rating

All too often ignored is the ripple-current rating of the output capacitor. The probable reason this rating may be

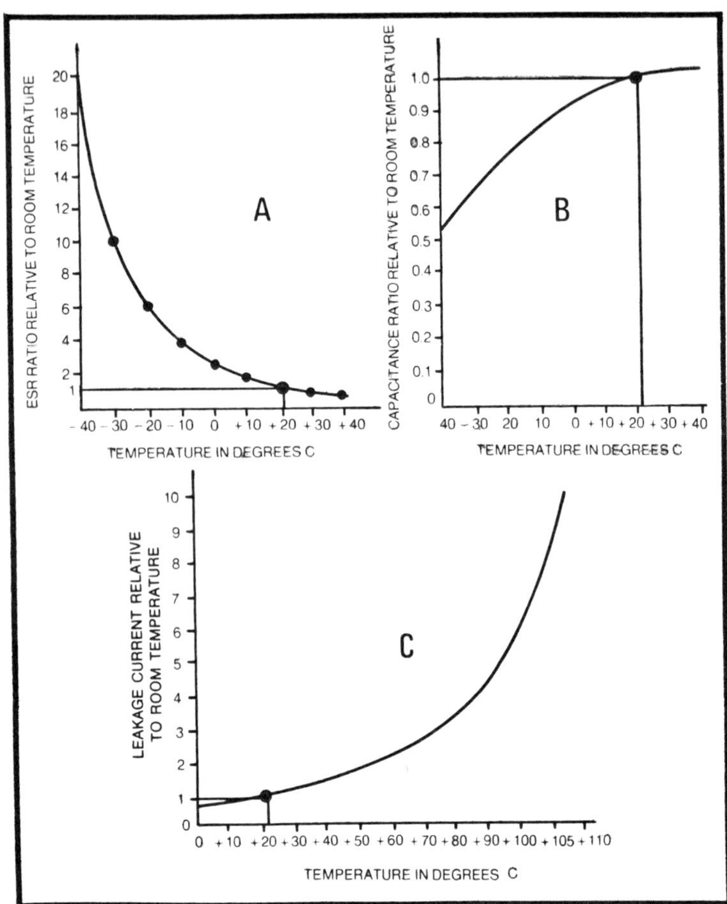

Fig. 7-4. Typical temperature dependencies in electrolytic filter capacitors. (A) Equivalent series resistance as function of temperature. ESR is normalized at unity for room temperature (21°C). Capacitor is a 1500 μF "computer grade" aluminum electrolytic. (B) Capacitance as function of temperature. Capacitance is normalized at unity for room temperature. (C) Dc leakage current as function of temperature. Leakage current is normalized at unity for 21°C.

overlooked is because it has not been of great importance in the linear regulator. In a switching-type power supply, however, the lifespan of the output capacitor is bound to be adversely affected if its ripple-current capability is not adequate for the job. Fortunately, all major capacitor makers can provide ripple-current ratings. Our task is to determine the rating required, and there is a general rule of thumb method that has proved to be quite reliable. Although somewhat empirical, this method is predicated on sound logic. The following assumptions apply:

- The waveshape of the current variation in the inductor is a sawtooth.
- This current constitutes the only ripple to which the output capacitor is subjected.
- A typical peak-to-peak value of this sawtooth current is 20% maximum direct current (I_L) delivered to the load. This is based upon the generally encountered guidance to design, which suggests making the inductor current increase and decrease 10% with respect to the DC component of the inductor.

As a consequence of the foregoing, the ripple current (I_R) in the output capacitor can be computed from the formula:

$$I_R \qquad \frac{0.2I_L}{2\sqrt{3}}$$

The factor in the denominator is introduced to convert the peak-to-peak current wave into its RMS value. The relationship can be further simplified to $I_R = 0.058I_L$. Generally, it is wise to incorporate a margin of safety after this computation has been made, to ascertain that the manufacturer's rating corresponds to the appropriate ripple frequency of the switching frequency of the power supply.

The ripple current generates internal heat in the capacitor, with the attendant changes in temperature-dependent parameters, such as those shown in Fig. 7-4. Not shown, because of its somewhat nebulous nature, is the relationship of life to internal temperature, as elevated temperatures greatly reduce the life expectancy of any electrochemical component. It has often proved difficult to insure a moderate ambient temperature for capacitors, much less to aggravate the situation by permitting excessive ripple currents. Even with an appropriately rated capacitor, dangerous internal temperatures can develop when there is no provision for heat removal from the external surface of the case.

Selecting a Capacitor

There are two basic families of electrolytic capacitors from which to choose: aluminum and tantalum. The aluminum types are available in many quality grades and fabrication techniques. The tantalum types comprise distinct subtypes: foil, solid, and wet slug. At first glance, these might present a bewildering array with regard to selection. The old philosophy from the days of unregulated supplies, and carried over in linear regulators, was to search out capacitors having the highest figure of merit, defined by (capacitance $\times$ voltage)/cost. Obviously, this resulted in getting the most for the least.

However, the concept of getting the highest rated voltage and capacitance for the least cost can be very misleading when we are working with switching-type power supplies. As has been discussed, the impedance characteristics, primarily ESR, must be taken into account if the full potential of the switcher is to be realized. With line-frequency, thyristor-type supplies, cheap aluminum electrolytics of dubious quality can often be used, but all too often this leads to noise generation, poor reliability, and generalized criticisms of all switching-type power supplies. Good-quality aluminum electrolytics probably provide the best compromise, if one must be made, between cost and performance. These comprise "computer grade" types and other types made especially for switchers. Some of the specialized constructions, however, have very compelling features, such as very low ESR and ESL, rather than low cost.

The outstanding feature of all tantalum types is their high volumetric efficiency; which is to say that a lot of capacitance can be stuffed into a given volume. The wet-slug tantalum has been the undisputed champion in the capacitance-versus-volume contest; and in applications where size and weight assume great importance, such tantalums naturally merit attention. Solid tantalums appeal where there is great emphasis on longevity—both shelf life and operating life—but they have not been generally available in either large capacitance or high voltage values, as have other types. The foil-type tantalum is a good capacitor for switchers, but is not cost-competitive with aluminum types.

It is true that good capacitors have played a big role in making the switcher a technological and commercial success. But it is also true that the present success of switching-type supplies has inspired the capacitor makers to devise even better-suited types. Because of this impetus, it is difficult to say that one type is inherently better than another. At one point in time, and for a given application, one type may offer the optimum mix of parameters per dollar.

Where switching rates are very high and current demand is low, nonelectrolytic capacitors will more frequently be found as adjuncts to the previously discussed electrolytic types, providing bypassing at high frequencies where the ESL of electrolytics becomes too high.

Whatever type of capacitor is selected, successful use depends greatly upon packaging and wiring techniques. For example, the ESL requirements of a capacitor can be reduced through careful wiring procedures. Short, wide straps are the best, and even these can be paralleled to further reduce self-inductance. It would be self-defeating to buy a four-lead capacitor, with elegant impedance characteristics, and then connect it into the circuit using a foot or so of conventional wire. More and more, the art of designing switching-type supplies is incorporating radio-frequency techniques.

THE INDUCTOR

Trying to make an ideal component, such as a perfect solid-state switch, is an unattainable goal, but it focuses our efforts in the right direction. Thus, in the case of the switching transistor, we know that we must emulate as closely as possible a switch which is a perfect conductor during *on* time, a perfect insulator during *off* time, and one that changes its conductive state instantaneously. Pursuing a similar philosophy, what would our idealized inductor be like? The needs of the switching-type power supply indicate that such an inductor would be magnetically unsaturable, would have zero DC resistance, and would have no stray capacitance in its windings. This immediately suggests an air-core inductor, but for practical inductance values this would require too many turns of fine wire, which would also increase interwinding capacitance. The capacitance is undesirable because it promotes both series and parallel resonances that degrade circuit operation. (At the very least, we would hope that the first resonant frequency of our inductor would be much higher than the switching frequency.)

The only way to reduce the resistance of an air-core inductor is to use large wire which leads to an oversized monster. In order to cut down the number of turns needed, we

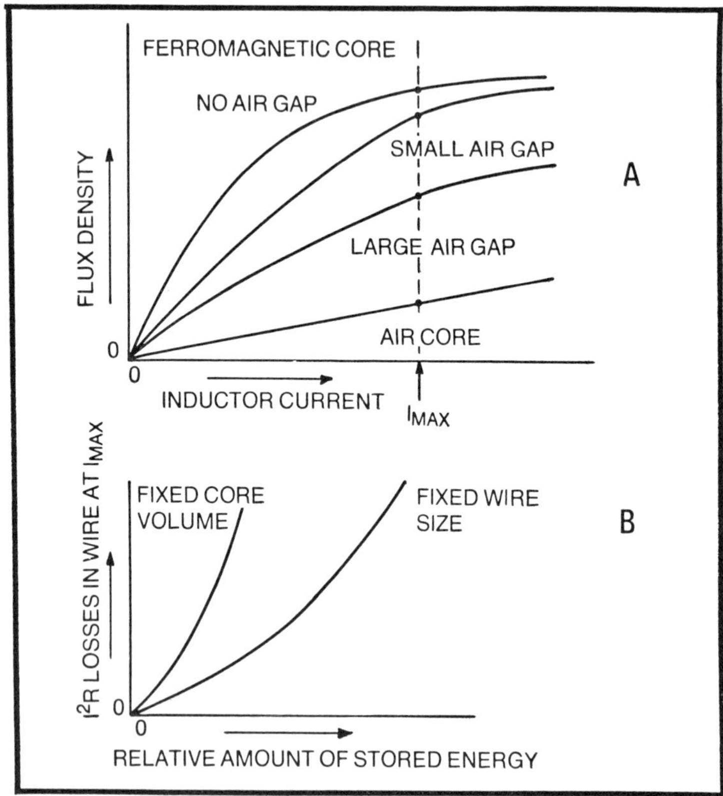

Fig. 7-5. Many tradeoffs are involved in the design of a good inductor. (A) A linear magnetization curve is very desirable, but is found only in oversized, air-core inductors; a compromise is to use a ferromagnetic core with a large air gap. (B) Too large an air gap requires too many turns of wire, which increases I^2R losses; a good rule-of-thumb compromise is to limit such wire losses to about one percent of total power-supply output power.

must include a ferromagnetic core. And, to avoid the effects of premature and especially abrupt saturation of the core, an air gap must be provided. The optimum but practical inductor is then a product that results from tradeoffs involving wire size, number of turns, effective magnetic-core permeability, and air gap size. Inasmuch as all of these quantities are interrelated, the design of a suitable inductor is as much an art as a science. And, as might be anticipated in such circumstances, more than one approach can be used. The magnetization and loss curves of Fig. 7-5 show the nature of the conflicting parameters involved in the design of a practical inductor.

Core Materials and Structures

It is convenient to use *ferrite cup cores* because the wire may then be easily wound on a bobbin and inserted into the core. Additionally, it is not difficult to establish a suitable air gap empirically in order to linearize the magnetization curve. Another benefit provided by the air gap is a relatively gentle passage into saturation when large currents are passed through the windings.

Molybdenum Permalloy *toroids* (ring-shaped cores) also make excellent cores for these inductors. Of course, there is some sacrifice in winding convenience compared to bobbin windings, but it fortunately happens that inductors in switching-type power supplies, and especially those operating at the higher frequencies, require relatively few turns of heavy wire compared to the monstrous "chokes" used in other supplies. The necessary air gap is actually present in this material, but it is effectively distributed throughout its volume in ferromagnetic particles that are separated by nonmagnetic binder material. As a result, the manufacturer has good control over the effective air gap in toroidal cores, and they are available in various A_L values (A_L being the inductance in millihenrys for 1000 turns of wire). Cores with relatively low A_L values tend to make optimum inductors for switching-type supplies, since a low A_L corresponds to a large effective air gap. Powered iron toroids have overlapping characteristics with Permalloy toroids with regard to the requirements of switching-type supplies, and thereby merit competitive consideration.

Ferrite toroids are a possibility, but the saturation characteristics are not as easily controlled, and this leads to design headaches. The emphasis in ferrites has been primarily directed to making higher A_L values rather than lower ones. The manufacturers of ferrite material have, however, made available an intriguing selection of core structures which appear to be application-oriented towards the needs of switchers, since their two-piece construction permits the inserting of air gaps.

Other core types have been more or less successfully used. These include laminated structures and air cores for specialized high-frequency designs. Laminated cores suffer from relatively high hysteresis and eddy-current losses at ultrasonic switching rates.

The least that can be expected from an inductor which enters saturation too rapidly is a change in the switching rate of a self-oscillating regulator. To a greater or lesser degree, depending upon other factors, this may increase ripple voltage and worsen DC regulation. A manifestation not quite so benign is the catastrophic destruction of the switching transistor, or the free-wheeling diode. After all, when the energy stored in the inductor is required to abruptly change, the deficit or excess energy must be either quickly supplied or absorbed, which tends to produce high, often destructive, peak currents in the switching transistor and free-wheeling diode. Consequently, it rarely pays to skimp on the design or cost of an inductor when you consider the cost and reliability of other associated components that it affects.

8

Semiconductor Switching Components

The semiconductor components that we discuss in this chapter are the discrete devices which perform the switching operation within the power supply. Switching transistors and diodes generally perform the DC voltage switching in switching regulators, inverters, and converters. Triacs and SCRs perform a similar function in switching AC voltages. There are, however, interesting marriages of these components in many systems, so that it cannot be said that any one device is limited to a specific switching operation. But it can be said that a knowledge of these devices is essential, since their optimum operation and survival in a switching-type supply places great emphasis on their ability to function as a truly ideal switch.

SWITCHING TRANSISTORS

There are many types of switching transistors available, and these run the gamut from modernized germaniums to sophisticated monolithic Darlingtons. Often, more than one type will suffice for the job at hand. In general, the choice is narrowed by such considerations as whether an NPN or PNP type is required, and whether strong consideration must be given to high voltage, high switching speed, high current, or cost.

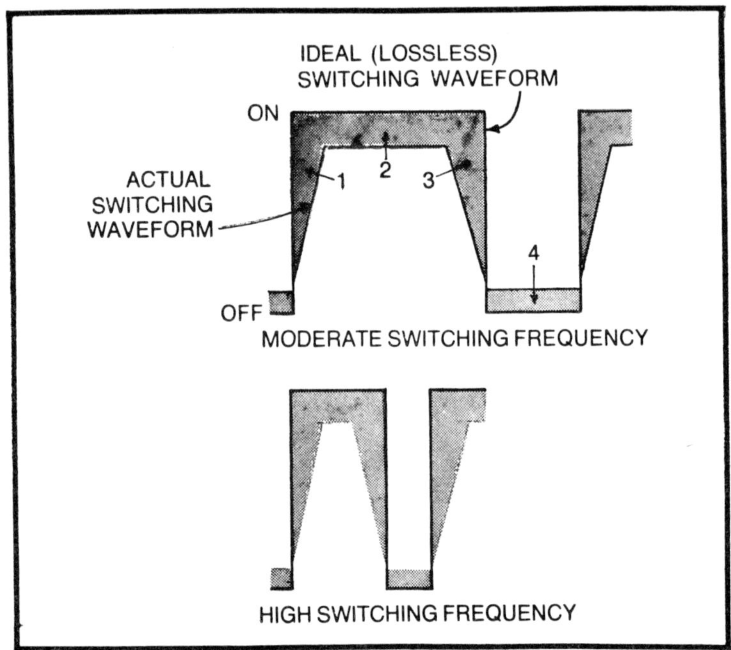

Fig. 8-1. A real switching transistor dissipates power, whereas an ideal switch would not. Power is dissipated in four regions: (1) during the turnon transition, (2) during the on time, (3) during the turnoff transition, and (4) during the off time. At high switching frequencies, most of the power dissipation occurs during turnon and turnoff transitions.

Initially, one has to be aware of the basic differences between an ideal switch and an actual switching transistor. In Fig. 8-1, a segment of an ideal switching waveform is illustrated; together with it is the actual waveform that a real switching transistor would produce. The perfect rectangular wave is a fantasy—one to be approached but never attained. The beauty of such a perfect switching waveform is that losses are zero *all the time*. The ideal switch is either all the way *on* or all the way *off*, and the transition between the two states is instantaneous. When this ideal switch is *on*, it delivers full current with no voltage drop across its terminals; when this ideal switch is *off*, it withstands the full voltage across its terminals and passes no current. The trapezoidal wave, the actual switching waveform, is something we must tolerate in a real transistor, but we expend much time and effort urging it to become more like the ideal.

The actual switching waveform has four regions of power loss. When this real switch is *off*, it is not entirely *off*; it allows a small leakage current to flow to the load. Because the emitter–collector voltage is highest during the *off* interval, this loss, though often low, can still be significant. When the real switch is in its *on* state, it again is not entirely *on*; the collector–emitter saturation voltage, $V_{CE(sat)}$, is sustained across its collector–emitter terminals. Although this voltage may be relatively low, it occurs just when the collector current is highest, and therefore this loss is also very significant.

The shaded areas of Fig. 8-1 reveal yet another deviation from the ideal. The actual switch requires time to make *on* and *off* transitions. During such transition times, the switch is operating as a linear element, a class A amplifier; it is dissipating power because it has appreciable voltage across it, as well as appreciable current through it. Fortunately, the switch usually spends only a relatively short time in these transition regions; but if the switching frequency becomes too high, the transition power losses become dominant. Power dissipated during these transitions also contributes greatly to transistor failures, and for this reason manufacturers often provide graphs showing various combinations of voltage and current that constitute "safe operating areas" during switching transitions.

General Transistor Characteristics

Table 8-1 provides generalized guidance with regard to the commonly used transistor types and fabrications. Because of economics, availability, second-source requirements, and ignorance, strange transistors are sometimes found in strange places. With the exception of the Darlingtons, the trend of frequency capability increases as one proceeds from left to right in the table columns. However, excellent monolithic, triple-diffused Darlingtons are now in a state of accelerated technological development; these structures are not inherently relegated to second or third place in the matter of frequency capability. There is also considerable overlap between adjacent columns.

Manufacturers indicate frequency capability in two ways. A common method is to designate f_T, the frequency at which

Table 8-1. Power Transistor Application Guide.

	GERMANIUM	SILICON, SINGLE DIFFUSION	SILICON, EPITAXIAL BASE	SILICON, EPITAXIAL DOUBLE DIFFUSION	SILICON, TRIPLE DIFFUSION	SILICON DARLINGTON
POLARITY	PNP	NPN	NPN/PNP	NPN/PNP	NPN	NPN/PNP
FEATURES	Low cost. High current capability. Low collector/emitter saturation voltage.	Rugged—has good surge-absorption capability. Excellent safe operating area	Practical combination of ruggedness, frequency capability and voltage rating. Efficient switch for many workhorse applications.	Highest frequency capability of commonly available power transistors.	Second best for high frequency switching. By far the best for high voltage ratings.	Low drive requirements. Simplifies assembly.
POSSIBLE DISADVANTAGES	Requires close attention to thermal situation. Limited Collector-voltage ratings.	Limited frequency ratings.	May prove to be a marginal performer for sophisticated high-frequency switchers.	Rugedness is limited by low safe operating area.	Limited current ratings. safe operating area is fair to good.	Limited current ratings. Second-source suppliers are not always available.
APPLICATIONS	Inverters and linear reaultors.	Good for operating under abusive conditions, and for linear regulators.	A good switch for 20 kHz regulators, and for linear regulators.	Switchers, inverters, and converters working from 20 kHz to over 100 kHz.	High frequency switchers and inverters, especially those operating in the 200—1000V range.	General purpose switchers where inordinate demands are not made on current, voltage, or switching rate. Also good for linear regulators.

the current transfer ratio of the common-emitter configuration becomes unity; f_T is also known as the gain—bandwidth product of beta cutoff frequency. The gain declines with respect to increasing frequency at 6 dB per octave; knowing this, we can work backwards and establish the approximate beta cutoff frequency, defined as being 3 dB below the current gain (or beta) at a low frequency, such as 1000 Hz. The beta cutoff frequency should be at least ten times the switching frequency; this is predicated upon the assumption that it takes 10 harmonics to construct a fairly rectangular waveshape.

The preceding frequency evaluation may not be valid for switching rates above several kilohertz because storage phenomena, incurred in the *on* or saturation mode, tend to predominate. Therefore, transistor manufacturers often give data more relevant to operation in the switching mode, such as the delay, rise, storage, and fall times under specified operating conditions. From such information, one can plan the switching rate and duty cycle in a realistic way.

The triple-diffusion structure may be subdivided into the *annular* and the *etch-cut* variations. The annular structure tends to display voltage ratings up to 400 or 500 volts, and the f_T may be 30 MHz or so. The etch-cut provides a classic example of tradeoff, for its voltage capability approaches 1000V, and its f_T may be on the order of 10 MHz. The safe operating area reveals more electrical ruggedness for the high-voltage than for the high-frequency types.

The loss mechanisms depicted in Fig. 8-1 do not tell the whole story about dissipation in the switching transistor. As one might suspect, power is also dissipated in driving the transistor, because a strong drive is required to force the transistor well into collector saturation during the *on* period. However, this also tends to keep power dissipation from $V_{CE(sat)}$ low. Therefore, it would appear that selection would favor transistors with high current gain and low emitter saturation voltage.

To a considerable extent, every time we optimize a parameter in a transistor, we pay for it by the degradation of another parameter. If we postulate an overall figure of merit for transistors by multiplying together the numbers representing all of the significant parameters, this product would tend to remain approximately constant; for one parameter is increased at the expense of a proportionate decrease in some other parameter. For example, voltage capability tends to deprive the transistor of frequency capability. From such a vantage point, advances in transistor technology can be defined as processing or fabricating techniques which result in improvement of all or most of the significant performance characteristics.

Switching Considerations

Driving a switching transistor is not just a matter of producing saturation during the *on* time. In Fig. 8-2, it is seen that the base current does not cease immediately when the transistor is supposed to be *off*. Because of charge storage in

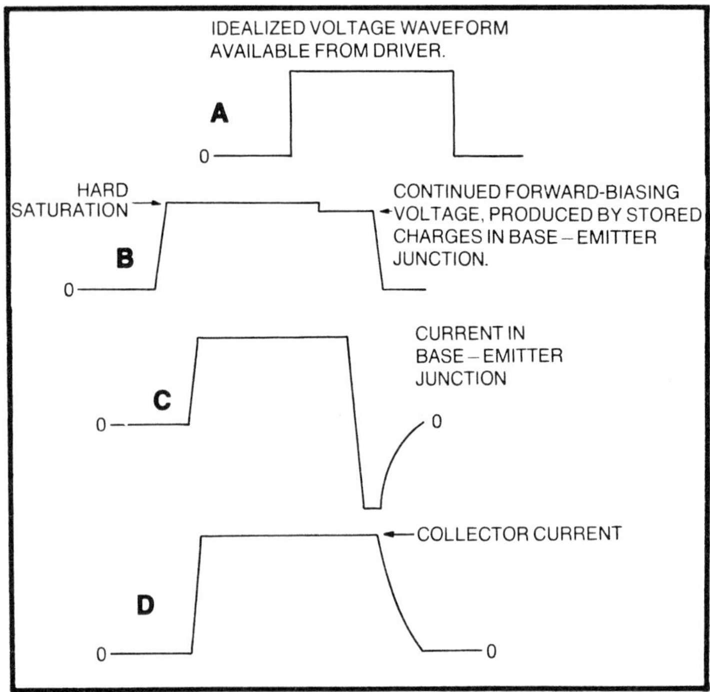

Fig. 8-2. Effect of base-charge storage on fall time of voltage at collector. (A) Ideal drive voltage delivered to base of switching transistor. (B) Voltage waveform actually monitored at the base. (C) The base current. (D) The current wave at the collector, which is the current available to the external circuit (inductor) during on time.

the base–emitter junction, some time is required for the base current to decay to zero. During this time the collector current continues to flow as if the transistor were still driven from an external current source. This phenomenon largely governs the fall time of the collector current. On the other hand, the rise time of the collector current improves with increasing base drive current. Thus, the losses in the turnon and turnoff regions of Fig. 8-1 are affected oppositely by base drive. Since the base drive is intimately connected with the loss in the *on* region, it is readily seen that there must be an optimum value of turnon base drive. (This is further complicated by the fact that the base is often made negative during the *off* period in order to decrease losses at this time and to reduce turnoff time.) As might be expected, optimum base drive is frequently determined empirically on a prototype model in an endeavor to obtain the highest efficiency possible.

Because of parameter tradeoffs in transistors, there is no universally accepted type. A germanium transistor complies with the requirements of one switcher, whereas another requires a sophisticated, triple-diffused, silicon type. But there is one criterion all designers must abide by: the concept of *safe operating areas*, or SOA. The SOA imposes bounds on transistor operation *in addition to* that which would prevail from the standpoint of power dissipation alone. Stated simply, you cannot simultaneously apply maximum current, voltage, and power ratings. The concept of SOA is shown graphically in Fig. 8-3. If the transistor were a simple passive device, such as a 100-watt resistor, it generally would not matter if the 100W resulted from 10V and 10A from a DC source, or 50V pulses at 10A from a switched source with a 20% duty cycle. Not so with the transistor, for it is subject to collector current limitation, to secondary breakdown at much lower power dissipation than would result from mere extension of the curve depicting maximum allowable power dissipation, and to collector–emitter voltage limitation because of avalanche breakdown.

Figure 8-3, however, is not complete. With real transistors, a power/ temperature derating curve generally must also be applied. And it turns out that there is an allowable relaxation

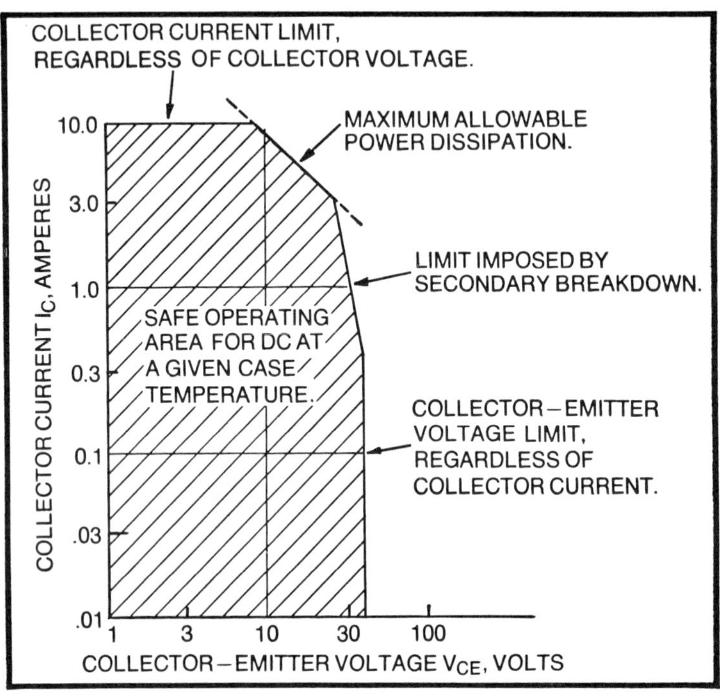

Fig. 8-3. Basic concept of the safe operating area of a power transistor. Note that the maximum allowable power dissipation, if extended, would indicate that any combination of collector current I_C and collector—emitter voltage V_{CE} would be satisfactory, as long as the product $V_{CE}I_C$ remained constant. This is not the case, for the power dissipation actually allowable over a practical operating range encounters power limitation from the secondary breakdown, the collector current limitation, and the collector—emitter voltage limitation.

of the boundaries of the SOA for switching waveforms. As might be expected, a more favorable use of the transistor's ratings becomes possible with shorter *on* times. Therefore, a more realistic approach to SOA is that shown in Fig. 8-4, from which allowance can be made not only for the basic bounds illustrated in Fig. 8-3, but for the way in which these bounds change with respect to duty cycle and to temperature. Thus, we need a family of SOA curves with *on* time as a parameter, and a power/temperature derating curve. Such information is usually provided by the major manufacturers of switching transistors, but it may not encompass all conditions of secondary breakdown; if in doubt, go to the manufacturer for more complete information.

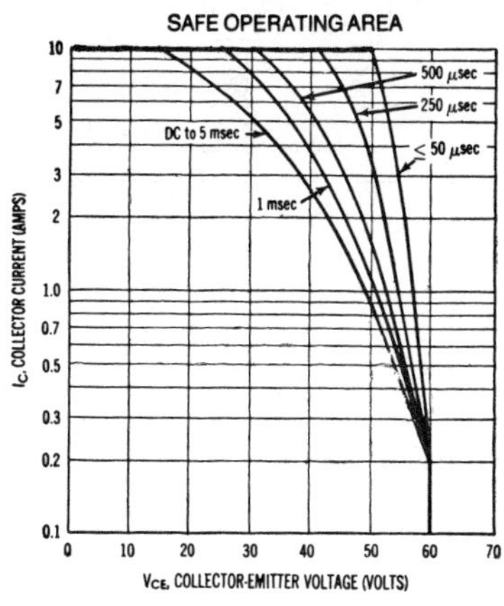

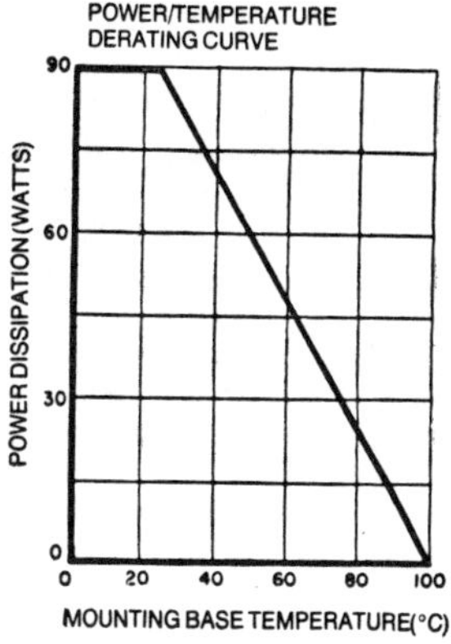

Fig. 8-4. Typical set of curves showing SOA and power/temperature derating information. When the mounting-base temperature exceeds 25°C, the data from the power/temperature curve imposes a further restriction on the product of collector current and collector—emitter voltage. Thus, the SOA curves may have to be adjusted accordingly. The power/temperature curve is shown as either a function of case temperature or collector junction temperature. In the latter case, a formula is generally provided to enable calculation of case temperature. (Courtesy of Motorola Semiconductor Products, Inc.)

Although we can estimate whether a switching transistor is "fast" or "slow" just from considerations of gain—bandwidth product f_T, a reliable indication of exact switching performance is not readily obtained in this manner. Therefore, transistor manufacturers also provide curves depicting delay, rise, storage, and fall times of those transistors likely to be used for switching. A typical set of such curves is shown in Fig. 8-5. By summing up delay, rise, storage, and fall times, we can more easily come to meaningful conclusions concerning allowable switching rates and duty cycles. The curves also enable one to estimate the losses likely to be incurred during rise and fall times. Other things permitting, the *on* time should be great enough to render rise time and fall time negligible in comparison.

DIODES

Diodes are commonly encountered in the circuits and systems of switching-type power supplies. We will confine our discussion to rectifiers and free-wheeling diodes; however, the

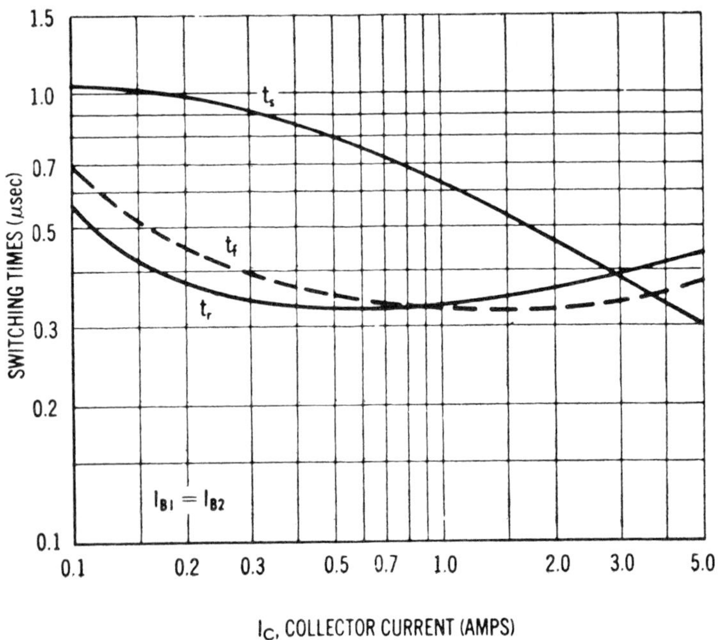

Fig. 8-5. Typical set of curves showing actual times governing switching speed. For a given collector current, this particular transistor cannot have an on time less than the sum of rise time t_r, storage time t_s, and fall time t_f. For high switching efficiency, the on time should be much greater than the sum of the rise and fall times. A fourth curve, turnon time delay t_d, is sometimes shown; its value is usually small, but it imposes a further limitation to switching rate. (Courtesy of Motorola Semiconductor Products, Inc.)

general ideas will be applicable to other circuit functions, such as commutation, holdoff, charging, and various others involving thyristors.

A common mistake made with early switching-supply designs was to select diodes according to their ability to withstand the worst-case conditions of forward and reverse voltage, forward current, temperature constraints, and surge and short-circuit current. This sounds like good, conservative, engineering design; and so it was, until the switching rate of these supplies began to increase. After several kilohertz was reached, the audible noise assumed a nuisance value—or worse, as higher power levels were reached. This provided impetus for a large jump in the frequency domain, to the

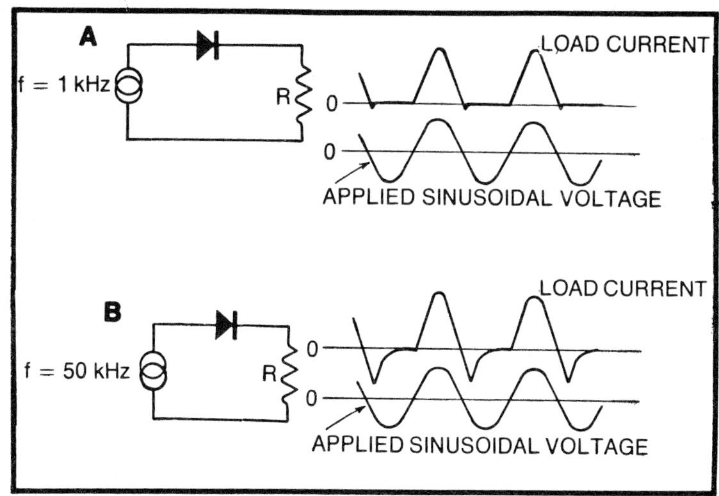

Fig. 8-6. Rectifying behavior of a typical diffused silicon diode. (A) Performance at 1 kHz: reverse-recovery characteristics are barely perceptible. (B) Performance at 50 kHz: reverse-recovery characteristics are a major feature of rectified current.

ultrasonic portion of the spectrum, starting with approximately 20 kHz. It was no longer sufficient to select diodes just because they had good rectifier characteristics at 60 Hz; we now had to be concerned with the *frequency capability* of diodes.

Real Diode Behavior

We have been deluged with the near-ideal characteristics of diffused-junction diodes, and such diodes are indeed quite remarkable. Their forward voltage drop is reasonably low for many applications, and little contention can be offered with regard to their very low leakage currents, even at high voltage levels. Nor can they be faulted on such matters as reliability or cost. In Fig. 8-6A, the half-wave rectification behavior of a large, diffused-junction silicon diode is about as ideal as one can expect at 1 kHz. But we do not see the losses from less than perfect conduction.

If, however, the frequency is increased to 50 kHz, the presence of the reverse-conduction region serves to drastically lower the rectification efficiency. The rectifier runs hotter and the overall efficiency of the power supply is reduced. As with switching transistors, the inability of the diode to faithfully turn off is caused by stored charges. The storage takes place during forward conduction; both the junction and the bulk material are affected and must be depleted of these charges. This does occur, but during the time required for charge "clean-out" by the reverse-voltage portion of the sine wave, excess power dissipation is taking place in the diode. And at the same time, the output voltage is lowered and the filter capacitors are subjected to increased ripple current. The rate of recovery from this postconduction behavior can be manipulated; best results are attained when the amount of absorbed charge is reduced as a consequence of processing and fabrication of the diode. Quick recovery, however, can generate other problems, such as RFI and circuit disturbances from ringing phenomena. Two diode manufacturing techniques have resulted in satisfactory high-frequency rectifying diodes.

In one technique, *fast-recovery* characteristics are produced, usually by gold doping the silicon semiconductor material. A significant improvement occurs, and substantially clean rectification is obtained at higher frequencies. As might be anticipated, such success is a tradeoff—adversely affected are the forward voltage drop and the reverse-leakage characteristics. In the overall picture of diode action, the tradeoff is a favorable one for high-frequency operation, since it is particularly desirable to eliminate or greatly attenuate reverse conduction stemming from charge storage.

The other method of dealing with charge storage is one of circumvention: a diode is made which does not make use of minority charge carriers, as there is virtually no charge to become involved in storage. Such a diode is called a Schottky barrier diode, otherwise known as a *hot-carrier* diode. Unlike a silicon junction diode, the Schottky fabrication utilizes a contact between a metal and semiconductor material, usually N-type. With appropriate processing, a unilateral conduction rather than an "ohmic" contact is produced. For practical purposes, this diode is free of storage phenomena, and its high-frequency performance is affected only by its internal capacitance, which is small enough to permit good, clean rectification beyond 100 kHz.

110

If the worst that could be anticipated from ordinary rectifier diodes in high-frequency circuits was the requirement for more effective heat removal from the diodes or somewhat lowered efficiency, perhaps there would have been less impetus for the development of high-frequency diodes. But, since circuit processes are interrelated, the poor performance of one component can adversely affect the operation of other components.

Switching Considerations

In Fig. 8-7A is shown a simplified high-frequency (50 kHz) converter. We can suppose that this is part of a switching-type regulated supply with a feedback path from the DC output to the inverter driver, which would control switching transistors Q_1 and Q_2 to vary the duty cycle of the inverter. But for the purpose at hand, consider only the 50 kHz rectangular wave that is being generated and rectified. Figures 8-7B and C then display the switching waveforms of good-quality diffused diodes and gold-doped fast-recovery diodes. It is evident that departure from ideal behavior is much greater with the conventional diodes than with the fast-recovery types.

Closer inspection reveals significant aspects of the distorted waveforms in Fig. 8-7B and D. In B, the diode current waveform shows considerable interference with the process of rectification. Among other things, much excess heat must be removed—this can be inferred by the large area of the wave beneath the zero axis. The painful consequences of the bad rectification appear in Fig. 8-7D, in which the switching transistors are subjected to a very high current peak—one enduring more than long enough to make its energy impact felt. These transistors are also forced to spend an appreciable fraction of *on* time in the unsaturated or linear region. It's a safe bet that the SOA will be exceeded by such behavior. Not shown are other possible adverse effects, such as higher ripple current through the output filter capacitor and increased output ripple.

Gold-Diffused vs Schottky Diodes

Fast-recovery diodes are processed with the aid of gold doping of the silicon semiconductor material, and by this

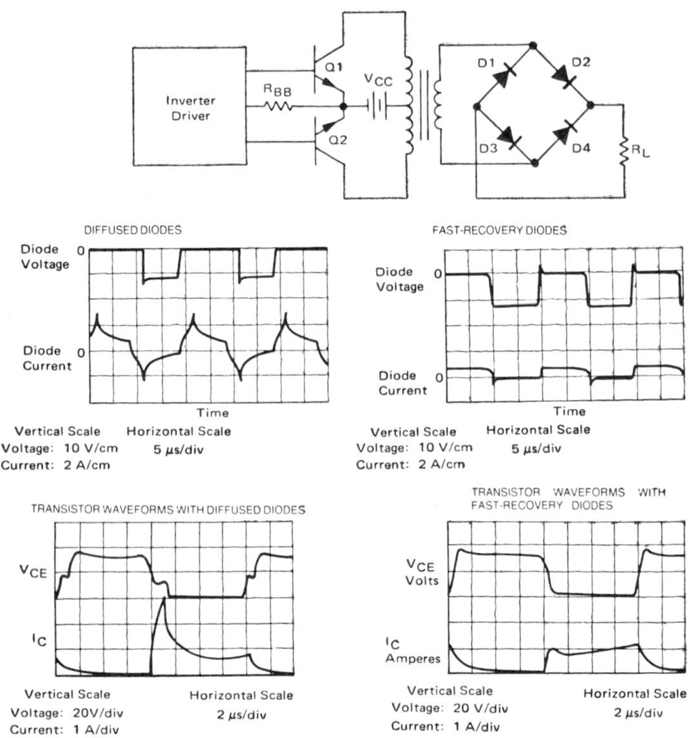

Fig. 8-7. Comparison of diffused and fast-recovery diodes when used in a 50 kHz DC/DC converter. (Courtesy Motorola Semiconductor Products, Inc.)

technique, very good reverse recovery characteristics are obtained at 100 kHz and beyond. The Schottky diode, being a majority-carrier device and inherently blessed with minimal charge-storage problems, also exhibits good reverse recovery well beyond 100 kHz. Is one diode better than the other? The excellent reverse recovery of the gold-diffused diode is obtained as a tradeoff with other parameters. One of these is the forward voltage drop. It happens that conventional silicon diodes are already in trouble when applied in low-voltage rectification systems. Because of computer technology, there is great emphasis on 5 or 6 volt supplies, often with considerable current capability. The forward voltage drop of the rectifiers constitutes a serious power loss at relatively low voltages. One helpful technique is to use a centertapped full-wave rectifier circuit rather than the bridge (which imposes the voltage drop of *two* diodes, worsening an already undesirable situation). Of course, a rectifier diode with lower forward voltage drop than silicon junction types must necessarily merit consideration, and this is where the Schottky diode shines.

Figure 8-8A shows the way forward voltage drops compare in Schottky and conventional silicon diodes of similar current capability. Not only is the Schottky diode a better choice for low-voltage, high-frequency rectification, but it remains the choice for low-voltage, low-frequency supplies because of its low forward drop. Its temperature capability, however, is generally inferior to silicon junction devices, but it is nevertheless superior to that of older germanium rectifier diodes. As a matter of fact, the Schottky diode is in its element in low-voltage applications because its reverse leakage current increases quite rapidly with higher reverse voltages. What technology will bring tomorrow is not too clear, but present Schottky diodes can approach excessive dissipation in the reverse region when the reverse voltage across it reaches about 30 volts, as shown in Fig. 8-8B; and the situation is more adverse at high operating temperatures. Much development work is being expended to extend this temperature limitation. Until recently, problems arose when the "junction" temperature hit 100°C, but currently available units are able to operate satisfactorily and reliably up to 125°C.

Fast-recovery, gold-diffused junction diodes become increasingly competitive as forward voltage drop diminishes in importance—that is, for the rectification of higher voltage, say about 10−15V. Where forward drop is not very important and the voltage capability of the Schottky suffices, it turns out that the gold-diffused diode may be a better choice in terms of cost.

SILICON CONTROLLED RECTIFIERS

SCRs, like switching transistors, are available in a wide variety of types. Indeed, the relevant characteristics for certain switching functions are not always easy to decipher from the specification sheets. The SCR is a device which is dependent upon *regenerative* circuit action. Once triggered,

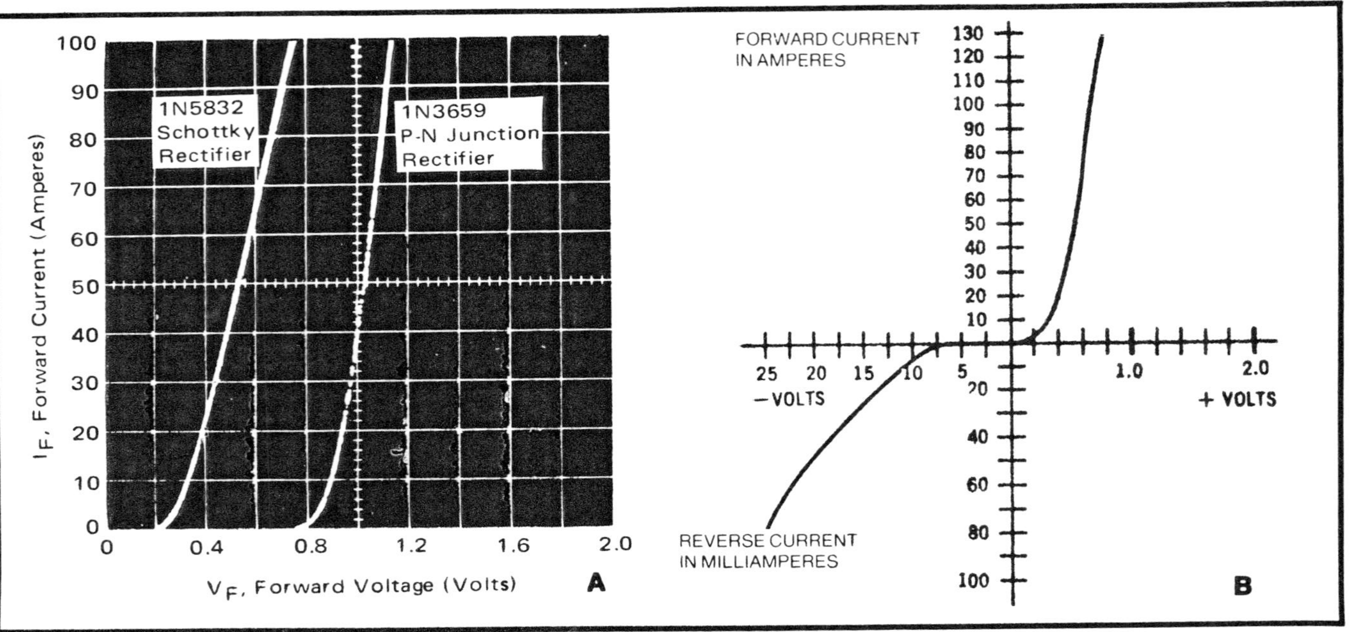

Fig. 8-8. Comparison of forward voltage drop in Schottky and conventional silicon diodes. The voltage drop of the Schottky rectifying diode is comparable to obsolete germanium rectifiers; however, the temperature characteristics of Schottky diodes make them superior devices.

the gate is deprived of control and a number of malfunctions can occur. A simple SCR phase-control circuit can be used as a preregulator and operated from the 60 Hz power line. The troubles likely to be encountered in such a circuit tend to be minimal. However, in higher-frequency applications and in circuits such as inverters, where turnoff is brought about by commutating techniques, it is possible to rapidly accumulate a box of burned-out thyristors without half trying!

Inexpensive SCRs are generally intended for the consumer market—for applications where the basic accomplishment is simple power control. These units are often made by the alloy-diffused process, so that gate-triggering voltage and current, anode–cathode blocking voltage, and holding current may have sloppy tolerances. It would be wrong to say that these relatively inexpensive devices cannot be used in switching-type power supplies, but to use them under the illusion that all SCRs are the same is an open invitation to trouble. Some of the malfunctions common to SCR switching circuits are:

- **Failure to turn off**—This can result from excessive junction temperature, inadequate blocking-voltage rating, transients in the main supply, or a poor trigger circuit. A likely cause with "bargain" devices is a low dv/dt rating; the rate of voltage change due to transients can turn such a device on, even though other turnon mechanisms are absent—catastrophic destruction is likely.

- **Spurious turnon**—This is caused by basically the same reasons as above. Again, inadequate dv/dt figures prominently in this malfunction, tending to damage the SCR.

- **Good operation but short life**—This is probable indication of inadequate di/dt capability. High initial current densities cause localized heating which gradually brings about deterioration and ultimate destruction. Thermal fatigue of solder-bonded pellet is also a possible cause of malfunction.

- **Latchup and commutation difficulties at high switching rates**—Charge storage imposes time delays in SCRs, just as it does in transistors. The ratings of many SCRs, even quality units, apply for 60 or 400 Hz operation. Units for higher-frequency applications must be so specified.

All-diffused SCRs (including epitaxial types) tend to be superior for switching-type supplies than do alloy-diffused devices. An additional consideration involves the matter of gate–cathode geometry. The edge-fired geometry has been extensively used, but it often leads to unsatisfactory dv/dt and di/dt capability, and so is not suitable for high-frequency operation. It is true that the edge-fired geometry requires relatively low gating current, but it is doubtful that this is a compelling feature in the overall picture. The initial cost advantage of edge-fired geometries has narrowed as more development has been directed towards the superior center-fired geometry. There are many versions of this new geometry, but all derive their superiority over the edge-fired technique from the fact that more cathode area is in conduction per unit time.

SCRs, in which the pellet is either soft- or hard-soldered, can have poor reliability in switching-type power supplies, certainly in those with marginal heat removal. A much better manufacturing technique is compression-bonded encapsulation, in which a spring system replaces the solder bond. For large supplies, controlling currents in the tens and hundreds of amperes, the *disc package* provides excellent thermal characteristics because the pellet is heatsinked from *both* sides.

Switching Considerations

Thyristors enable the control of load power by interrupting the flow of load current, so they are similar in many respects to our basic switching circuit of Fig. 8-9. The thyristors used most often in switching-type power supplies are the SCR and the triac, although other PNPN devices are also important for triggering and timing. The transistor operates as a *driven* switch—with its conductivity at all times dependent upon actuating drive. The thyristor operates as a *regenerative switch* triggered from its *off* to its *on* state by a short-duration

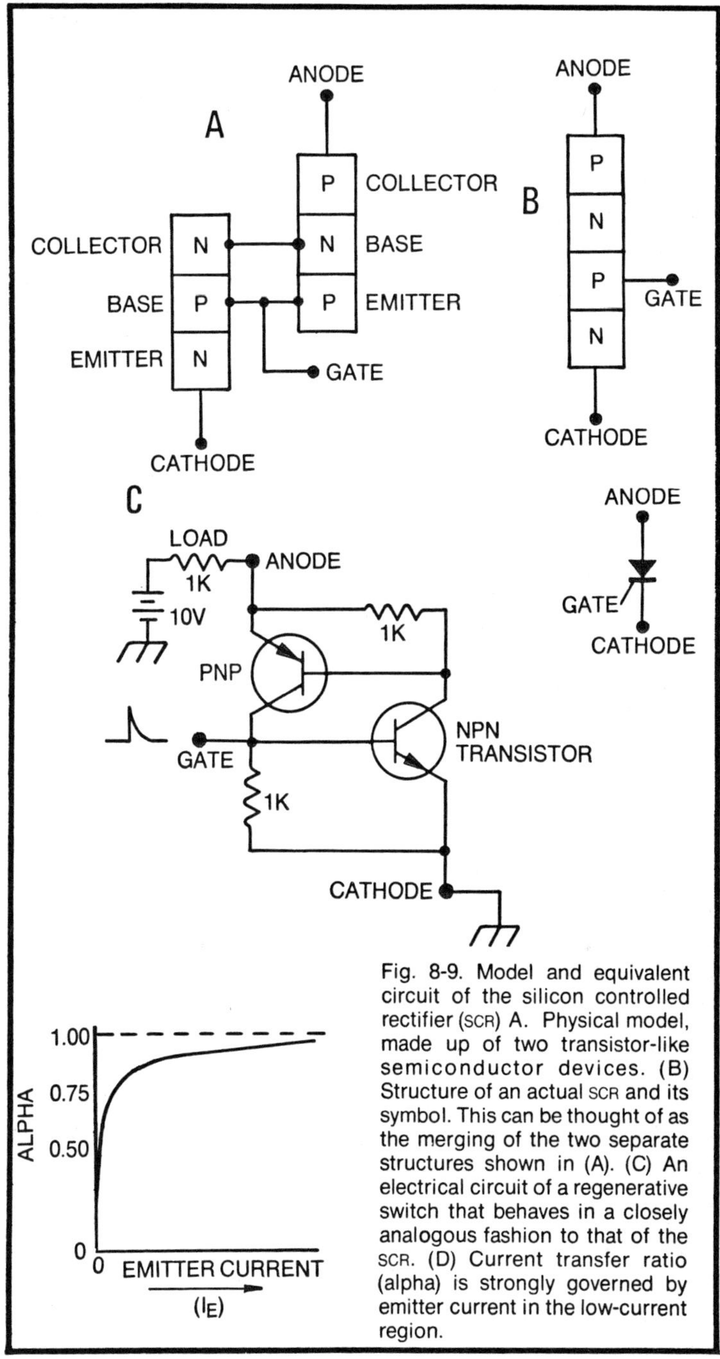

Fig. 8-9. Model and equivalent circuit of the silicon controlled rectifier (SCR) A. Physical model, made up of two transistor-like semiconductor devices. (B) Structure of an actual SCR and its symbol. This can be thought of as the merging of the two separate structures shown in (A). (C) An electrical circuit of a regenerative switch that behaves in a closely analogous fashion to that of the SCR. (D) Current transfer ratio (alpha) is strongly governed by emitter current in the low-current region.

pulse. Once triggered, the thyristor supplies its own gate drive and very rapidly goes into saturation, thereafter, the gate loses its control entirely. It does not matter if subsequent trigger pulses are delivered to the gate or if the gate is disconnected from the circuit. Once the thyristor is in its *on* state, it is latched as a closed switch.

The thyristor remains *on* until the current through it is either interrupted or reduced below a relatively low value—the so-called *holding current*. The thyristor then reverts to its nonconductive or *off* state. Having thus been reset, it is again receptive to a gate trigger pulse. Inasmuch as the simple thyristor circuit is fed from an AC source, load current is determined by *timing the occurrence of gate triggers*. In this way, more or less of each half-wave cycle is allowed to pass to the load. This method is known as phase control, but the effect is essentially that of duty-cycle or pulse-width modulation.

Figure 8-9 shows physical and circuit models of the silicon controlled rectifier. The two-transistor configuration closely duplicates the actual operation of the SCR. In essence, we have a latching binary, or flip-flop, in which one state change (*off* to *on*) is provoked by a gate trigger pulse, and the alternate change of state (*on* to *off*) is brought about by the next zero crossing of the AC input wave. (There are other modes of thyristor operation, but the described sequence of events is typical.)

The useful behavior of the SCR stems directly from the fact that *alpha*—the current transfer "figure of merit" in transistors—is a function of emitter current. Alpha is simply the ratio of output (collector) current to input (emitter) current with the transistor in the common-base circuit. The concept of alpha can be used in equations describing relationships in transistor circuits other than the common-base configuration, as in the dual-transistor circuit of Fig. 8-9. Since neither transistor has any forward-biasing arrangement, the stage is in its nonconductive state. Although a leakage current (I_{CO}) provides some forward-bias current to the base—emitter junction of the NPN transistor, the alpha developed by this transistor at such low currents is too low to start any chain of events in the overall circuitry. Thus, this transistor, and the PNP transistor also, is "dead."

The injection of a short-duration positive trigger into the "gate" terminal provokes a drastic change in the circuit. The NPN transistor has its alpha momentarily increased and thereby its collector current. This also serves to increase the forward-bias current of the NPN transistor, bringing it to life. The resultant collector current of the PNP transistor reinforces the forward base–emitter bias of the NPN transistor, *whether or not the gating pulse still exists*. Once initiated, both transistors are speedily transferred into their conductive states. Since one transistor reinforces the *on* state of the other, the entire circuit *latches* in the *on* state, allowing full current to be delivered to the load. In this latched state, trigger pulses have no further effect, and even a negative pulse applied to the gate will usually have no effect. To restore the circuit to its *off* state, the current must be momentarily interrupted either at the cathode or anode.

There are thyristors (e.g., triacs) with gate *turnoff* capability, but conventional SCRs are most often used as phase-controlled switches in power supplies, regulators, motor controllers, and other applications where power levels exceed several watts. Even at much lower powers, the conventional SCR is generally used.

The equation describing current flow through the transistors is

$$I = \frac{I_{CO}}{1 - (\alpha_1 + \alpha_2)}$$

where I = current flowing through the load
  I_{CO} = leakage current through the circuit where the transistors are in the *off* states of conduction
  α_1 = alpha of PNP transistor
  α_2 = alphs of NPN transistor

This equation states that the load current increases substantially as the *sum* of the two individual alphas approaches unity, at which point the load current is determined entirely by the circuit's own resistance. The transistor circuit then behaves as a closed switch. Of course, neither this circuit nor the thyristor structure it simulates has zero resistance or zero voltage drop in its *on* state, but since both transistors are in hard saturation, the voltage drop across the anode/cathode terminals of the device is relatively low. Active or "hot" transistors are those in which alpha closely approaches unity. But the foregoing equation tells us that even relatively poor junction devices, those with alphas considerably less than unity, can participate in the regenerative process which requires only that the *sum* of the current transfer ratios attain unity.

Voltage/Current Characteristics

From what has been said so far about the SCR, it might well be supposed that the consequence of applying an AC voltage to it would be essentially nothing. The rationale could be that, *providing no gate signal was applied*, the two-transistor model would not permit anything greater than leakage current to pass. When the polarity of the anode–cathode voltage was proper for normal conduction, no conduction would occur because of the lack of emitter–base bias on both transistors. For reverse polarity of the applied voltage, one would expect no action anyway. This reasoning is valid, but only if the applied AC voltage is kept within bounds. One of these bounds is familiar to us, for if the reverse voltage is high enough, avalanche breakdown will occur. In Fig. 8-10, reverse breakdown occurs at point A, and is designated as BV_R. When the SCR is in a phase-controlled switching circuit, it is important to avoid reverse breakdown, as this results in loss of both rectification and load control.

In Fig. 8-10, we see another breakdown at point B, designated as BV_F. It is very important to understand that this breakdown, for forward polarity of the applied AC wave, occurs *without any gate signal*. This may appear strange because the phenomenon resembles the action we expect from an initiating gate pulse—the SCR regeneratively switches to its *on* state and latches there. We are thus up against a basic limiting factor in circuit operation. What happens is that the leakage current increases enough to forward bias the emitter–base junction of the NPN transistor. This increase is

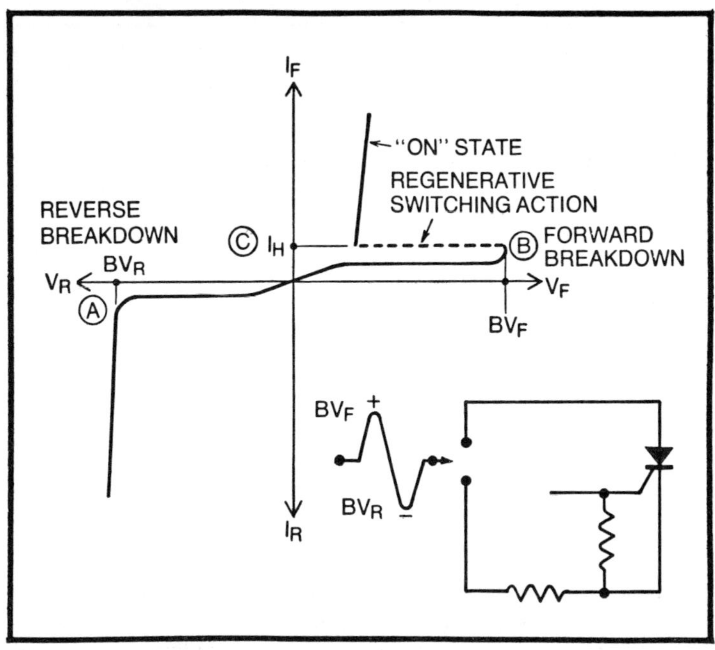

Fig. 8-10. Voltage-current characteristics of the SCR with zero gate voltage. Region A is similar to avalanche or zener breakdown in ordinary silicon diodes. This region must be avoided when the SCR is used as a phase-controlled switch. Region B must also be avoided when the SCR is used as a phase-controlled switch. It is true that the regenerative switching action must be produced, but it must be provoked by a gate signal, rather than by the forward breakdown voltage, BV_F. Region C indicates the holding current, I_H, and is the minimum anode current needed to sustain the SCR in its ON state of conduction.

sufficient to induce both transistors into a state of saturation. Obviously, point B must be avoided if we wish to control the SCR only by the timing of gate pulses. In Fig. 8-10, since it was desired to display the total dynamic characteristics of the SCR, the impressed voltage was deliberately made higher than would be proper for actual circuit operation. Because of the effects of temperature, as well as the rather sloppy tolerances on many SCRs, it is wise to incorporate a healthy safety margin to assure operation free from either of these breakdown phenomena.

In Fig. 8-11, the applied voltage is well within bounds, so neither forward nor reverse breakdown occurs. A triggering network is normally provided so that gate-induced turnon can

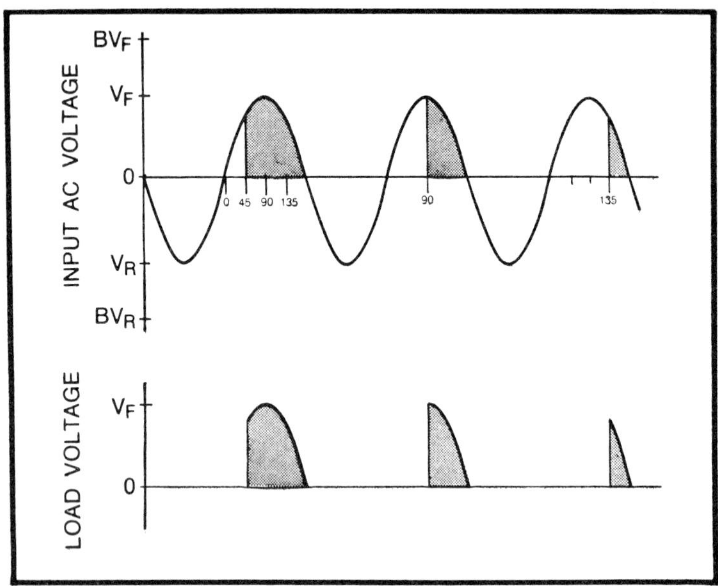

Fig. 8-11. SCR switching waveforms. The duty cycle or conduction angle is governed by the gate-triggering time of the SCR. The three examples illustrate 45, 90, and 135 degree conduction angles, measured from the positive-going transition of the input sine wave. The earlier the SCR is triggered into conduction, the more power is delivered to the load. Note that the peak amplitude of the AC voltage must not exceed the limits imposed by the breakdown voltages, BV_F and BV_R.

be adjusted. The triggering circuit could turn on the SCR immediately, at the positive-going transition of the input AC voltage, or the triggering could be delayed for a time. Figure 8-11 depicts several triggering delays. The shaded areas represent the portion of the cycle in which the SCR is in its *on* state, delivering power to the load. Simple triggering circuits merely sense the AC input voltage and trigger the SCR when some preset voltage level is attained; others utilize resistor—capacitor circuits which permit delays to be greater than 90°; and more sophisticated triggering circuits use feedback to regulate the output power.

Originally, SCRs were used almost exclusively for 60 Hz applications. Later, they performed efficiently in 400 Hz three-phase systems. At the present, there are SCRs available for use in inverters operating at frequencies beyond 25 kHz. The maximum frequency of the SCR is determined by the response characteristics of the semiconductor junctions. In a manner of speaking, the difficulties stem from *excessive* rather than attenuated response. As we increase the frequency of the applied voltage, we find an increased tendency toward turnon, *irrespective of gate signal*. The inherent capacitance existing across the junctions of the device can cause high-frequency turnon by allowing sufficient turnon current to be passed to the gate from the main anode voltage source. This is known as the dv/dt effect, being appropriately named because current through this capacitance is proportional to the frequency of the applied AC voltage.

Another performance parameter which must be kept within bounds is the di/dt value. This is the *initial* rate at which anode–cathode current increases following turnon. Lifespan and reliability are adversely affected if this rate is excessive, particularly in large devices. Often, di/dt can be slowed down by means of a small amount of series inductance. Since it may not be easy or practical to obtain timely and meaningful data, it is often wise to derate current capability. It is always a step in the right direction to provide for good heat removal. SCRs are available in different constructions, gate geometries, and packages. Consultation with a reputable manufacturer should always be the prelude to design, and even replacement, for the state-of-the-art for these devices progresses faster than their external appearance might suggest.

The gate-turnon characteristics of SCRs (and triacs as well) play an important role in both the turnon time and the amount of dissipation that the SCRs incur during the turnon transition. Manufacturers give both minimum and maximum values for gate-turnon voltages and currents. While it might seem desirable to select gate-drive parameters that favor the *minimum* side of the specifications, this results in slower turnon times which inevitably lead to increased dissipation, particularly at higher operating frequencies. It is true that increased drive values also increase the amount of dissipation in the gate circuit of the SCR, but it is the *total* amount of power dissipated in the SCR that is most important. Consequently, it is actually easier on the SCR to provide near-maximum values of gate-triggering pulses.

SCRs and Triacs

It is commonly stated in technical literature that the triac is a full-wave thyristor, one that permits passage of both alternations of the AC wave to the load. This sometimes leads to a misconception, for although the SCR operates as a half-wave rectifier, it does not follow that the triac functions as a full-wave rectifier. In most switching-type power supplies, the ultimate output is regulated direct current. Of course, this does not necessarily exclude the triac, for rectification can always be imposed following the control function. A more natural preference for the SCR stems from its rectifying property, and the fact that triacs are more limited in voltage, current, and power ratings. From these considerations, there should be no surprise that the triac has not displaced the SCR in switching-type power supplies.

In Fig. 8-12 is shown the basic arrangement used for phase control of full-wave DC power with two SCRs operating in conjunction with a pair of conventional rectifying diodes. (This basic technique is utilized in the switching supply illustrated in Fig. 3-3.) An interesting method of accomplishing the same objective with a *single* SCR is depicted in Fig. 8-13. The

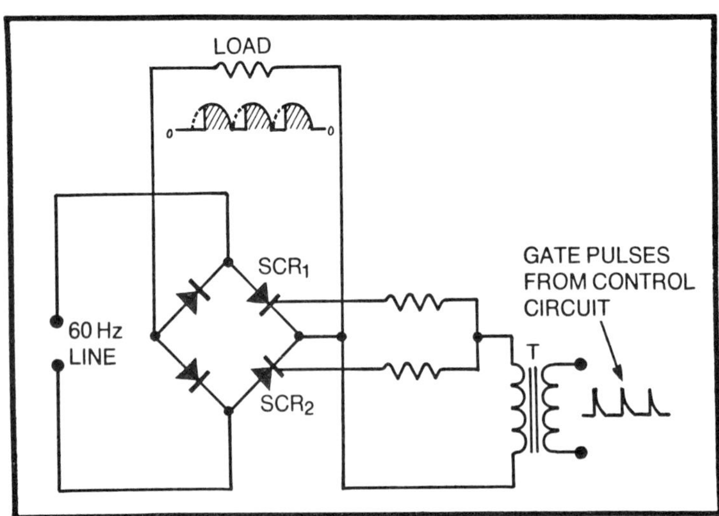

Fig. 8-12. Switching circuit for full-wave power control by SCRs. The circuit is essentially a modified rectifying bridge in which two conventional rectifiers are displaced by SCRs.

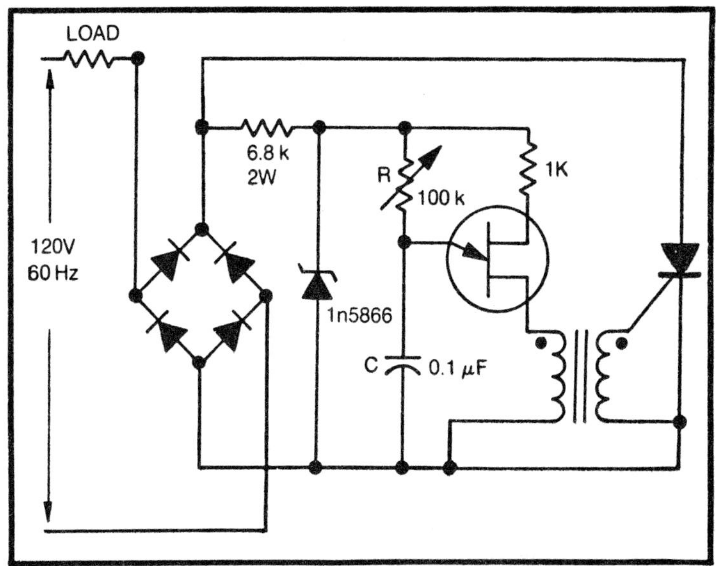

Fig. 8-13. Alternate method of scr full-wave power control. In this arrangement, a single scr produces a phase-controlled output for both alternations of the ac cycle.

gate-firing circuit uses a unijunction relaxation oscillator. For simplicity, the manually controlled version of this basic technique is presented—the power delivered to the load is adjustable by means of variable resistance R. (A wider control range can be obtained if the UJT circuitry is empowered from a small auxiliary DC supply, rather than from the main rectifying bridge as shown.) In a switching-type regulator, resistance R could be replaced by the collector–emitter circuit of a bipolar transistor; an output-derived DC error signal could then be applied to the base to vary the charging rate of capacitor C in such a way as to bring about regulation of the sampled output-voltage.

Whether this method is used in manual or in automatic controlled systems, the UJT commences an oscillation cycle at each zero crossing of the 60 Hz wave. This constitutes synchronization with the power line and is an essential operating feature. This feature depends upon feeding the UJT with *unfiltered* DC. Therefore, if an auxiliary supply is employed, as suggested above, there should be no filter capacitor.

Triacs are in general limited to low-frequency applications, although future types may overcome this disadvantage. Phase-controlled triac circuits are similar to SCR circuits, but benefit greatly from the fact that the triac requires only one gate signal, whereas an SCR circuit generally requires two. The gate of the triac can be triggered using DC voltages of either polarity, although the gate tends to be somewhat more sensitive when the gate voltage is of the same polarity as the anode. Since triacs are normally used only for AC switching applications, the worst-case turnon conditions must be accounted for in the design of the trigger circuits. As in the case of the SCR, it is always best to provide near-maximum gate drives, both to insure fast turnon and to reduce device dissipation.

Figure 8-14 illustrates a triac and its switching characteristics. The triac also exhibits both forward and reverse breakdown voltage, BV_F and BV_R. The conduction characteristics of a triac are symmetrical, and a gate pulse can turn on the triac with anode—cathode voltages of either polarity. (Nomenclature such as anode and cathode are meaningless with triacs, since the device is bidirectional in nature. Most manufacturers assign new terms for these electrodes—such as M1 and M2, for *main* polarity terminals. However, as long as we bear the bilateral nature of these electrodes in mind, we can continue to use the anode and cathode terms we are familiar with.) The triac is very much like two SCRs connected in reverse parallel, except for the triac's single gate terminal. The triac's behavior is otherwise equivalent to the SCR, and descriptions pertaining to the SCR are generally applicable to triacs as well.

THE OPTOISOLATOR

An interesting device, with considerable potential for use in switching-type power supplies, is the *optoisolator* (Fig. 8-15). Incorporating a light-emitting diode (LED) and a photodiode, phototransistor, or photo-Darlington light detector, this device provides the functions of level shifting and electrical isolation. Both of these functions give designers many a headache in switchers without a 60 Hz power transformer, or in those in which a high-voltage output

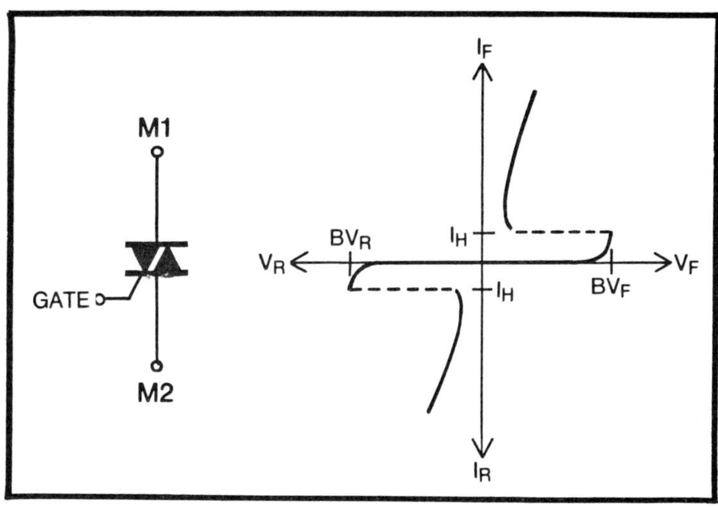

Fig. 8-14. The triac operates as a symmetrical AC switch. The unique feature of the triac is that it has only one gate terminal, which can be triggered by pulses of either polarity. For typical applications, the gate-trigger pulse must be of sufficient amplitude to reliably trigger the triac into conduction for either polarity of applied voltage.

is developed. Of course, it may be argued that this device is not really new, since packaged combinations of photoconductors, such as cadmium sulfide, in conjunction with incandescent lamps, have been around for many years. The performance and reliability of some of these older versions were notoriously poor—most engineers looked upon them as novelties, and there was no attempt to make them commonplace items. However, the modern optoisolator is truly different, for both the transmitter and receiver of the optoisolators have frequency capabilities more than adequate for the most sophisticated ultrasonic switcher. This contrasts dramatically with the slow response of incandescent lamps, neon bulbs, and photoconductive detectors made from various compounds of lead, selenium, cadmium, and sulfur.

Unfortunately, the predominant emphasis on the use of optoisolators has been for interfacing various logic circuits. The digital mode of operation will also find use in switching-type supplies, but the most immediate application appears to be in the switcher's feedback loop, where the optoisolator is used in its linear mode. The makers of these

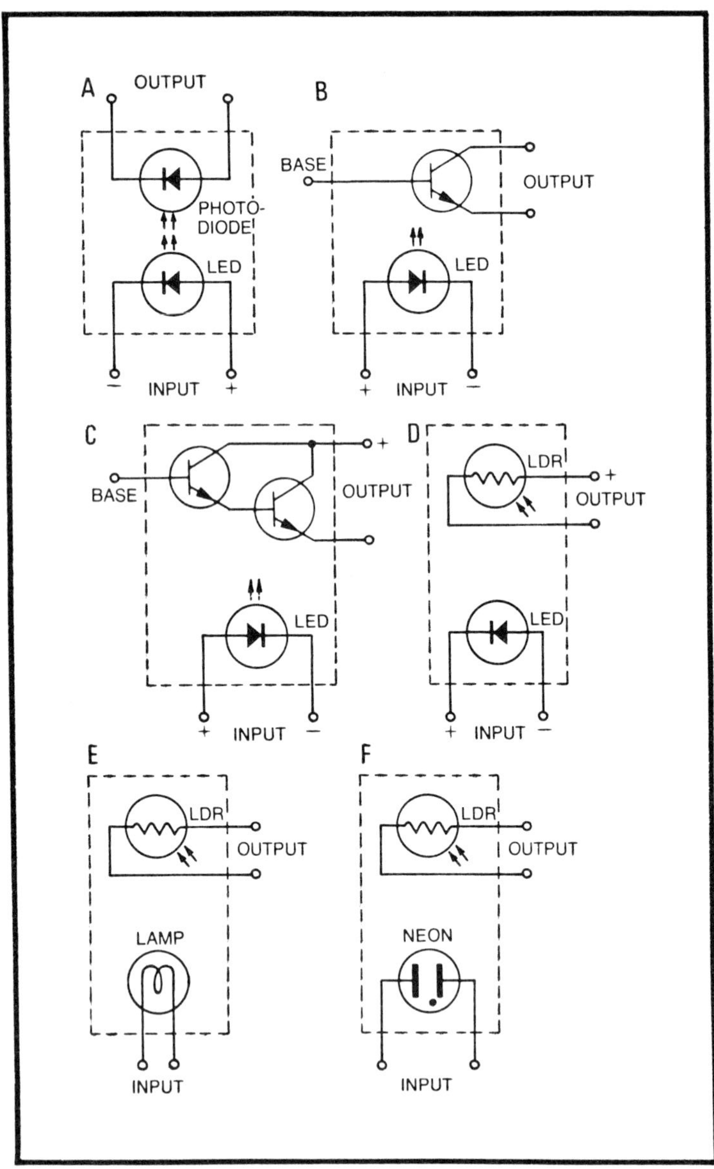

Fig. 8-15. Optoisolators come in many forms. Light-emitting diodes (LEDS) emit light when they are forward biased, as indicated in the figures above. In A the photodiode is operated in reversed-biased mode, since light acts to increase diode leakage currents. In phototransistors B and C, the collector−base junction of the transistor acts as a reverse-biased diode. Various forms of light-dependent resistors (LDRS) are also used in conjunction with different types of light-emitting devices (D, E, and F).

products have had some reservations about advocating such use, because the transfer function of the optoisolator is often not very linear. But the dynamic excursion of operating current need not extend into severely nonlinear regions, nor is the nonlinearity of great importance when the device is inserted in the negative feedback loop. The regulator derives its performance primarily from its error-signal canceling feature, and so it not adversely affected by nonlinearities, but tends to compensate for changes occurring within the feedback path. In practical resulting power supplies, optoisolators are easily incorporated using standard design procedures to insure stability in the feedback loop.

On the other hand, optoisolators tend to be quite temperature dependent, and their current-transfer ratio varies greatly among the various types. Those with photodiode output devices have the best response characteristics, but the output current is, at best, a small fraction of the input current. An optoisolator with phototransistor output can develop current-transfer ratios in the vicinity of unity, and when a photo-Darlington output is used, the current-transfer ratio can range up to ten. The latter type is also the slowest, but units are available with responses adequate for switching regulators. In all of these types, the current-transfer ratio is greatly influenced by the forward input current—another way of describing their nonlinearity. The base leads of phototransistors may be left unused or may be used for inhibit, strobe, or other control purposes. Also, by connecting a resistance from base to emitter, higher frequency response may be attained at the expense of sensitivity. Figure 8-16 shows the current-transfer curves for a typical optoisolator with LED input and phototransistor output.

Some types use a photoconductive output cell. This may be cadmium sulfide, cadmium selenide, or a similar material. The response time of such a cell should be considered before the device is used in a feedback loop, inasmuch as photoconductive devices tend to be slow. LEDs are usually infrared types, but the LED in Fig. 8-15D is often peaked in the visible region to achieve better overall efficiency than speed of response. Photoconductive types also require different circuit

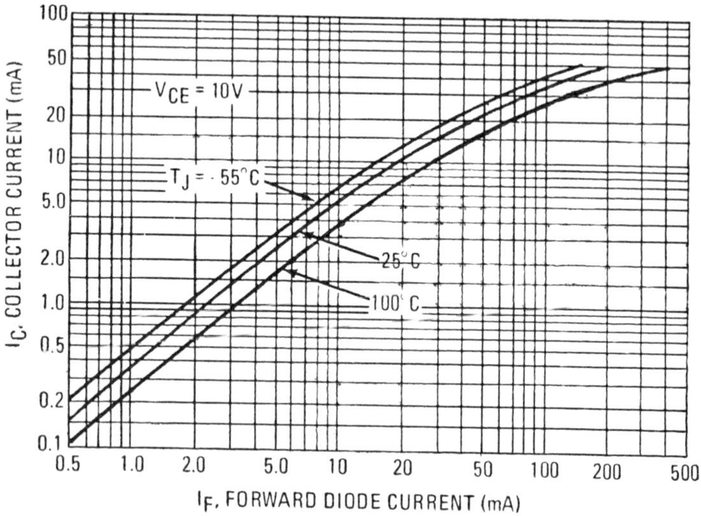

Fig. 8-16. Current-transfer characteristics of the Motorola 4N25 optoisolator. This device uses an LED and phototransistor and achieves a current-transfer ratio of about 0.5 for an input diode current of 10 mA. (Courtesy of Motorola Semiconductor Products, Inc.)

implementations, inasmuch as they generate no current or voltage, but rather a modulated resistance.

Optoisolators using incandescent or neon lamps are older versions. Long life spans can be obtained from incandescent lamps at very low currents. Telephone circuits have apparently achieved such reliability, but not all electronics designers have observed this precaution. An advantage of this type, as well at the neon-bulb version, is that an AC signal can be directly sensed. All optoisolators are enclosed in light-tight packages, and provide 1500V or 2000V insulation between input and output. Special designs extend this isolation to well over 10 kV.

9

Voltage References and Comparators

The regulation capabilities of a power supply depend largely upon the stability of the internal voltage reference source and the associated circuitry that is combined with it. While the ultimate accuracy of the output voltage depends upon the accuracy and temperature stability of the reference voltage, certain circuit functions—feedback-loop components, comparator or error amplifier characteristics—contribute to the overall stability and performance of the power supply. This is true of both linear and switching-type power supplies.

VOLTAGE REFERENCE SOURCES

The voltage reference source encountered most is the zener diode. Considerations of cost, reliability, and ease of implementation have been determining factors. Thus, in a large proportion of regulator schematics, we can quickly locate the zener diode reference source. Actually, the simple zener diode has shortcomings for applications where stability, temperature drift, and dynamic impedance are of importance. In the better quality linear regulators, these matters assume the utmost importance, and the voltage-reference source is often modified. In switching-type regulators, emphasis on tight regulation and temperature immunity has only recently received emphasis, because other inadequacies in the switching process did not make it worthwhile to pursue such goals. With modern components and circuit techniques, switchers are often designed to serve in applications that were previously the domain of the linear regulator. In such instances, it does not always suffice to simply throw in a cheap zener diode with a composition resistor.

As with the linear regulator, better results can be obtained with a *voltage-reference* diode. In reality, this "diode" comprises two or more diodes—usually one is a reverse-breakdown (zener) diode and the others are normal forward-conducting diodes. The diodes are connected in series so that their opposite temperature coefficients tend to cancel; zener diodes above 5V have *positive* temperature coefficients, while forward-biased junctions exhibit *negative* temperature coefficients of approximately -2 mV/°C. Although a 5.6V zener diode might have a positive temperature coefficient of $+2$ mV/°C, a 7.5V unit will tend to have a positive temperature coefficient of about 4 mV/°C. In the former case, a single forward-conducting diode would be used in an attempt to approach a zero temperature coefficient; in the latter case, two forward-conducting diodes would be necessary. This technique is not so simple as it might initially appear—either for the maker or the user. It is not at all easy to mass-produce diodes with the required uniformity to yield specific voltages and temperature coefficients when combined. The user must provide a constant-current drive, rather than a simple dropping resistor, if he is to advantageously employ the device. Together, these result in high cost; but temperature coefficients in the vicinity of $\pm 0.005\%$ are attainable.

Actually, zener diodes were long ago misnamed. It so happens that the *reverse-breakdown* phenomenon derived from two mechanisms: *field emission* and *avalanche*. Field emission (*zener* breakdown) predominates at the lower end of the voltage range, whereas avalanching is the important breakdown mechanism near 10V. Within this voltage range, *both* breakdown mechanisms are operative. Breakdown by field emission is actually a *tunneling* process, wherein charge carriers have a certain probability of jumping a forbidden zone. Such a breakdown mode has an exponentially described behavior, and its current—voltage characteristic is not very abrupt. Such a characteristic is undesirable because the actual breakdown voltage is not clearly defined. The dynamic impedance of such a gradual transition tends to be high—again undesirable in a voltage reference. A high dynamic impedance implies that a small change in current through the device will result in a relatively high change in voltage developed across its terminals; the lower the dynamic impedance, the better the bypass action for AC components. In this respect, a voltage reference with low dynamic impedance simulates a very large capacitor, but with the advantage that the impedance to AC is very low for *all* frequencies.

The avalanche-breakdown mechanism occurs when charge carriers in the junction are accelerated to a high velocity by an electric field. These carriers then impact valence-bound electrons which impart their kinetic energy to other electrons. The phenomenon is *cumulative*, and is manifested by a *very abrupt* increase in current through the device. (The action is similar to the ionization of a gaseous diode.) Recently, voltage-breakdown diodes have been introduced on the market in which breakdown in the 4—10V range is predominantly by avalanching. These low-voltage avalanche (LVA) diodes display exceedingly sharp breakdown characteristics. Microamperes rather than milliamperes are required for their operation, thus greatly reducing self-heating. The tremendous improvement stems from processing techniques. The improved devices are, among other things, usually fabricated by the epitaxial rather than the diffusion process. Unfortunately, these superior breakdown diodes are not depicted differently on schematics than ordinary types.

The reverse-breakdown characteristics for conventional and LVA 5.1V diodes are compared in Fig. 9-1, and clearly show the difference in the transition region of the diodes. The almost vertical slope in the conductive region of the LVA diode a superior voltage reference for regulators, but its abrupt turnon characteristic and low leakage current can enhance the performance of peripheral equipment as well. For example, SCR *crowbar* circuits are often used with both linear and

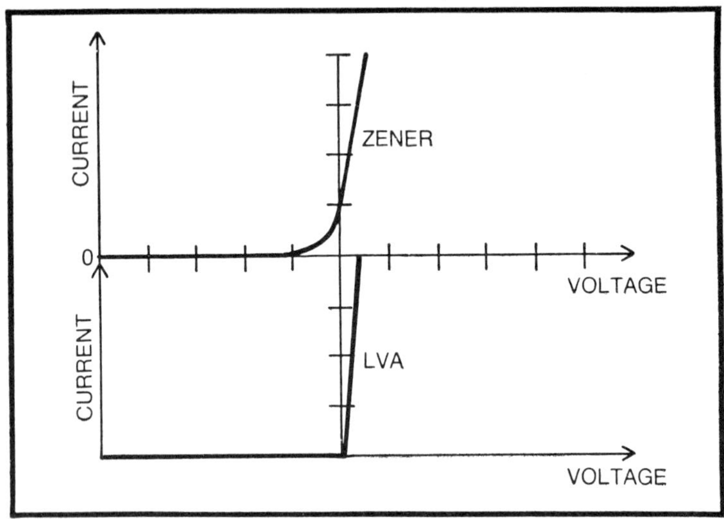

Fig. 9-1. Comparison of conventional zener diode and low-voltage avalanche diode. The slow, upward bending of the conventional diode waveform indicates a high forward impedance, while the sharp, almost vertical slope of the LVA indicates a much lower impedance.

switching-type power supplies to protect the load from overvoltage in the event of regulator malfunction. By using the LVA diode in the gate circuit of the SCR, the triggering can be made more precise than with conventional breakdown diodes. Not only can better protection be imparted to the load, but enhanced immunity to spurious operation from noise will generally result.

FET CONSTANT-CURRENT DEVICES

The field-effect transistor is inherently a constant-current device, and its current—voltage characteristic is similar to the pentode electron tube. In the simplest situation, as depicted in Fig. 9-2, only two terminals are available to the external world, and the device is commonly referred to as a diode. The characteristics of this simulated diode are also shown in the figure, from which it is seen that a substantially constant current prevails over a large portion of the operating region. If a precision resistor is connected in series with one terminal of the device, as in Fig. 9-3A, a constant voltage will be developed across the resistor and may be used as a reference voltage

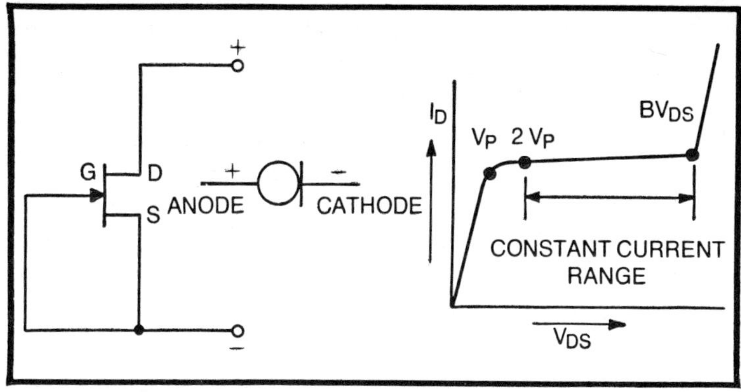

Fig. 9-2. The constant-current diode is actually a field-effect transistor with gate and source terminals connected together to form the cathode of the diode; the drain terminal then becomes the anode of the diode. Most constant-current diodes are made from N-channel FET; if P-channel devices were used, the polarities of the drain and source terminals would be opposite.

with the provision that the sampling device has an input impedance greatly exceeding the value of the series resistor. In regulated power supplies, this requisite is readily met by either a comparator or by a buffer stage.

Instead of using conventional FETS, better results are attainable from specially designed field-effect diodes that are

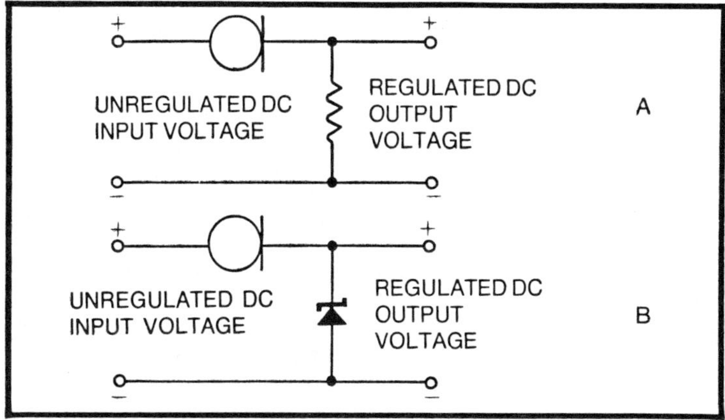

Fig. 9-3. Two applications of the current regulating diode, or CRD. The circuit in A provides an adjustable output voltage reference by using the constant voltage drop across a precision resistor due to the constant current flowing through it. The circuit in B provides an excellent voltage reference by biasing the reference diode at an optimum current level for obtaining a low temperature coefficient.

processed to optimize the constant-current characteristic. These devices have only two terminals inasmuch as there is an *internal* gate−source short. They are available in over thirty current increments from about 200 microamperes to 5 milliamperes. The minimum operating potential is very low—on the order of 1−3V, and the forward-breakdown voltage is often as high as 100V. Thus, these devices are adaptable to a wide variety of circuit situations. An advantage of using these specially made current-regulating diodes (CRDs) is that a zero temperature coefficient can be achieved at the factory by associating the CRD with a resistor of equal but opposite temperature coefficient. This usually prevails for units rated in the vicinity of 0.5 mA.

An excellent application for the CRD is in conjunction with a zener, LVA, or voltage-reference diode, as shown in Fig. 9-3B. A temperature coefficient of 0.001% over the 0−100°C temperature range may be attained with this technique. For optimum results, both the CRD and the voltage-reference diode should be specified for zero temperature coefficient in the vicinity of 0.5 mA.

There is more than meets the eye in the combination of CRD and zener. The CRD in the constant-current source exhibits an extremely high impedance to AC. The contrary is true of a zener or other voltage-reference source, in which an important figure of merit involves its very low AC impedance. When the two types of devices are combined in the manner shown in Fig. 9-3B, a unique low-pass filter is formed with a cutoff frequency near zero. Such a configuration theoretically offers very high attenuation to *all* AC frequencies. In practice, because of stray parameters, the circuit exhibits attenuation approaching 100 dB for frequencies up to several hundred kilohertz. Thus, most ripple and noise components riding on the unregulated supply line are effectively removed. In the simple circuit of Fig. 9-2, the output impedance is equal to $1/g_{oss}$ where g_{oss}, or an equivalent, is a commonly specified parameter. In this circuit, the constant current is equal to I_{DSS}, also a commonly specified parameter. The modified circuit of Fig. 9-4A provides any constant-current value up to the value of I_{DSS}. Additionally, output impedance increases as the regulated current is reduced by increasing R, because of the feedback action. The cascaded arrangement of Fig. 9-4B is capable of producing much tighter current regulation and much higher output impedance than is attainable from single-device circuits. For this circuit to operate properly, Q_2 must have a higher I_{DSS} than Q_1, preferably at least ten times greater. And it is important that both FETs are provided with drain–source voltages at least twice their pinchoff voltage, V_P, also a commonly specified parameter of FETs. This criterion for V_P actually applies to the single-device circuits as well.

The CRD can be advantageously used as part of a voltage-divider network for sampling the output voltage of the switching regulator. Such a scheme is shown in Fig. 9-5. When so utilized, the error signal is not attenuated, as is the case with the conventional potentiometric sampling divider. An example of such an application may be found in the advanced switching regulator design of Fig. 13-15.

SYNTHESIZED LOW-VOLTAGE REFERENCE

One of the techniques sometimes employed to obtain a reference voltage *below* that provided by zener diodes is to use

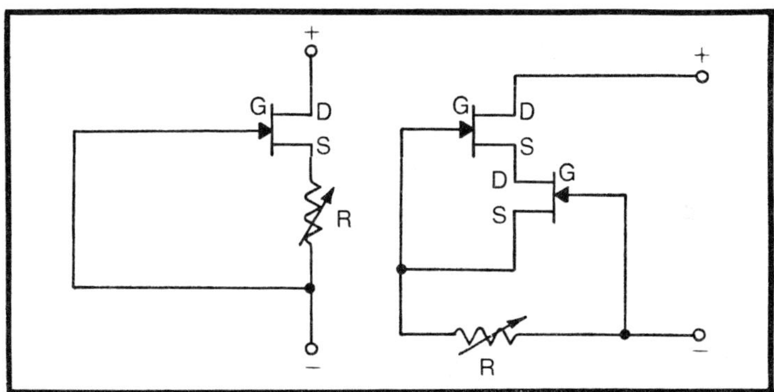

Fig. 9-4. Constant-current sources with adjustable current levels. The cascaded arrangement of two FETs offers much better regulation and higher dynamic impedance than the single-FET design. In both designs, however, increasing resistance R acts to decrease the value of constant current. Maximum current is obtained with R=0. For each individual FET device, there is an optimum value for R that will yield a nearly zero temperature coefficient. The manufacturer is able to alter the fabrication process to obtain a specific constant-current value and zero temperature coefficient with zero resistance.

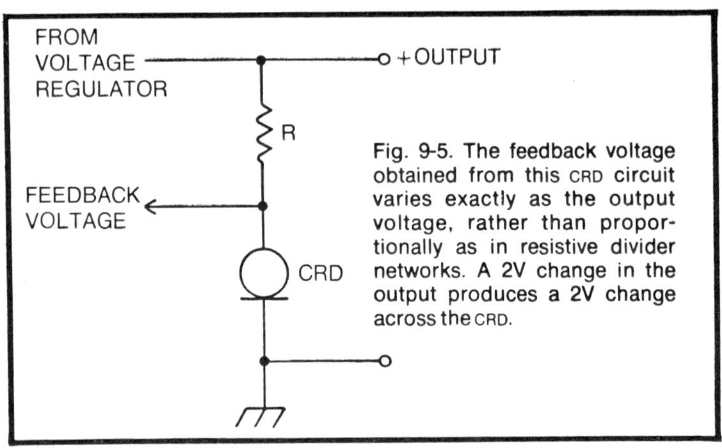

Fig. 9-5. The feedback voltage obtained from this CRD circuit varies exactly as the output voltage, rather than proportionally as in resistive divider networks. A 2V change in the output produces a 2V change across the CRD.

ordinary silicon PN diodes in their forward-conductance mode. Approximately 0.6V per diode is obtained in this manner. Thus, a 2.5V reference results from connecting four diodes in series. This method is not very satisfactory, since the regulation is quite sloppy, dynamic impedance is high, and reproducibility is not at all good. It may, therefore, be surprising that one of the finest low-voltage references is derived from the forward-conduction characteristics of PN junctions.

It has long been known that the relationship between the emitter—base voltage and collector current is highly predictable in transistors, and that two transistors in an unbalanced differential-amplifier circuit could be made to produce a current that was proportional to the absolute temperature. (A useful property for electronic thermometers.) A significant result occurs if the collector currents of the amplifier stages are maintained at a constant value, for it then happens that the base—emitter voltage of a third transistor (also operated with constant collector current) can be combined with the *base—emitter difference voltage* of the differential amplifier stages to produce a voltage which is *independent* of temperature. In other words, there is a certain way of summing the base—emitter voltage drops to obtain a reference voltage having a zero temperature coefficient. This occurs at the energy-band gap voltage of silicon, which is 1.205V.

It should be appreciated that the attainment of the zero temperature coefficient, together with other desirable properties of a voltage reference source, was never practical with discrete devices. Only the monolithic fabrication process, with its tight thermal coupling between devices and its excellent reproducibility of device characteristics, could be used to successfully translate theory into hardware. The result is a new series of devices that offer exceptionally low temperature coefficients down to the one-volt reference level, which can be used to much advantage over the typical 6.2−8.5V range of convential zener diodes. And since no breakdown mechanism is involved, noise generation is very low.

The schematic diagram of the LM113 is shown in Fig. 9-6. Transistors Q_1 and Q_2 form an unbalanced differential amplifier. A voltage, proportional to the *difference* of their base−emitter voltages, appears across R_4. To this differential

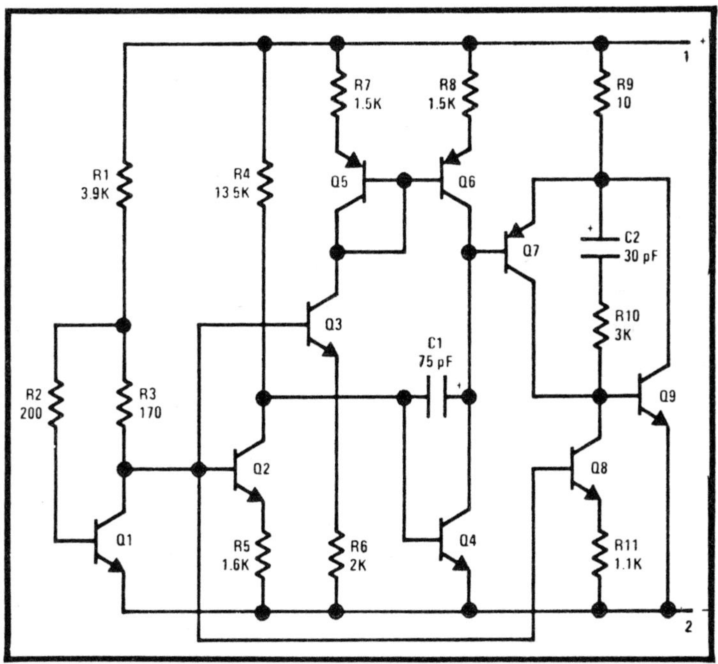

Fig. 9-6. Schematic circuit of the LM113. A precision 1.2V reference source. (Courtesy National Semiconductor Corp.)

base−emitter voltage is added the base−emitter voltage of Q_4 in order to develop the zero temperature coefficient, as previously described. Transistors Q_3, Q_5, and Q_7 provide a current-source load for Q_4. The remaining transistors are primarily involved in buffering the output and in producing a low dynamic impedance.

The characteristics of the LM113 are shown in Fig. 9-7. Note the near-vertical slope of the "breakdown" region,

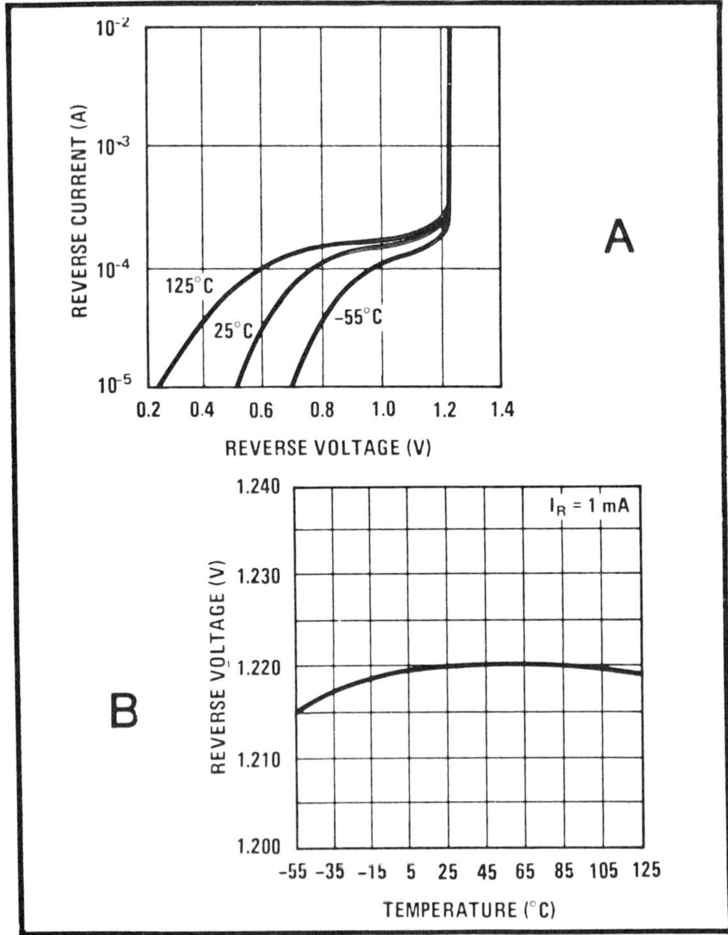

Fig. 9-7. Characteristics of the LM113 voltage reference source. The output reference voltage becomes very stable at forward currents above 1 mA. Note the expanded voltage scale used in depicting temperature effects. (Courtesy National Semiconductor Corp.)

inferring a low dynamic impedance. Although only the active-operating region of transistors are involved, the similarity to the characteristics associated with breakdown phenomena is quite remarkable. Indeed, the nearest conventional breakdown device—a 2.4V zener diode—is inferior in all respects as a voltage reference. The temperature characteristic attests to the validity of using opposing temperature coefficients within the IC to establish a near-zero temperature coefficient over a wide temperature range.

A 2V regulator using the LM113 "diode" is illustrated in Fig. 9-8. The LM113 is driven by an FET current source and serves as the voltage reference in a linear regulator involving an LM108 operational amplifier and a 2N2905 booster transistor. As shown, this circuit will operate on an input voltage as low as 3V.

THE COMPARATOR AND ERROR AMPLIFIER

This building block of regulatory circuits and systems has many names and more than one mode of operation. In the technical literature of servo systems, linear regulators, and switching-type supplies, we find such terminology as error amplifier, error-signal amplifier, sensing amplifier, sensor, comparator, differential amplifier, summing amplifier, etc. All of these terms are not necessarily synonyms; they encompass more than a single kind of circuit behavior, but they do possess a common feature no matter how named, their circuit function is to generate output signals which distinguish between the input conditions of "too high" and "too low." Inasmuch as these conditions pertain to the output voltage of the supply being regulated, the information contained in the response of the comparator (or its counterparts) can be utilized for *correcting* the very deviation responsible for such an error signal. Correction in the linear regulator occurs through the rheostat-like action of the pass transistor. In the switching-type supply, the correction results from an appropriate change in the duty cycle of the switching process.

In certain switchers, the comparator is very similar to those used in linear regulators; it may consist of a single

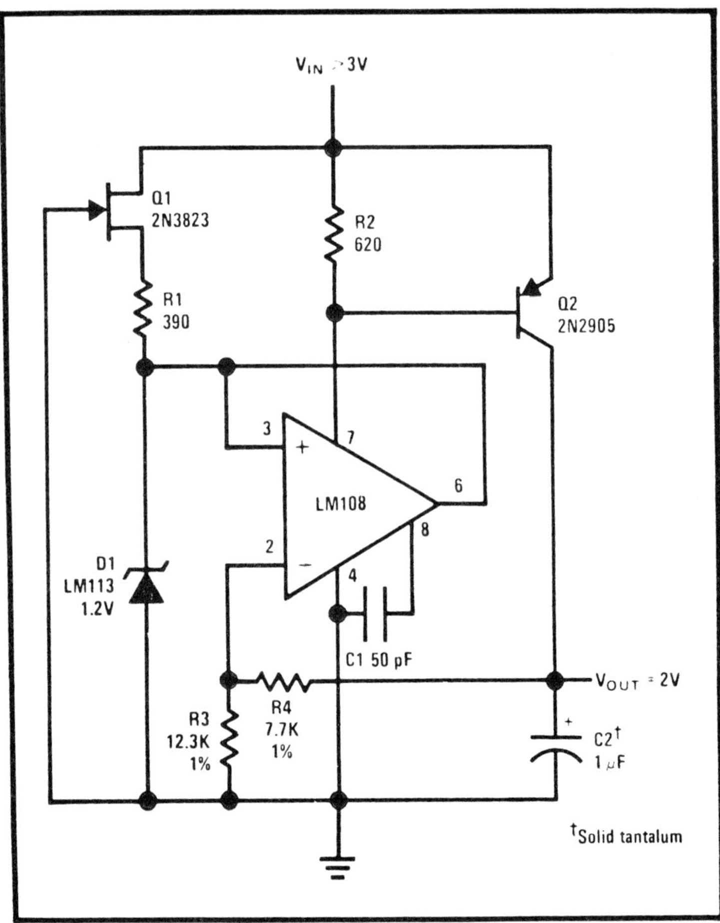

Fig. 9-8. A 2V regulator using the LM113 synthesized voltage reference source. In turn, this regulator would make an ideal voltage reference source for a switching-type regulator designed for low-voltage output. Note that the LM113, although a complex monolithic array, is depicted as a simple breakdown diode. (Courtesy National Semiconductor Corp.)

transistor, a differential pair of transistors, a complex array of discrete devices, an operational amplifier, or an IC module specifically intended for such functions. Examples are shown in Fig. 9-9. The use of monolithic ICs as comparators reduces the burdensome tasks of design and manufacturing. Other problems associated with imbalance, offsets, temperature tracking, and device tolerances can be neatly circumvented with ICs. The physical layout is invariably more conducive to

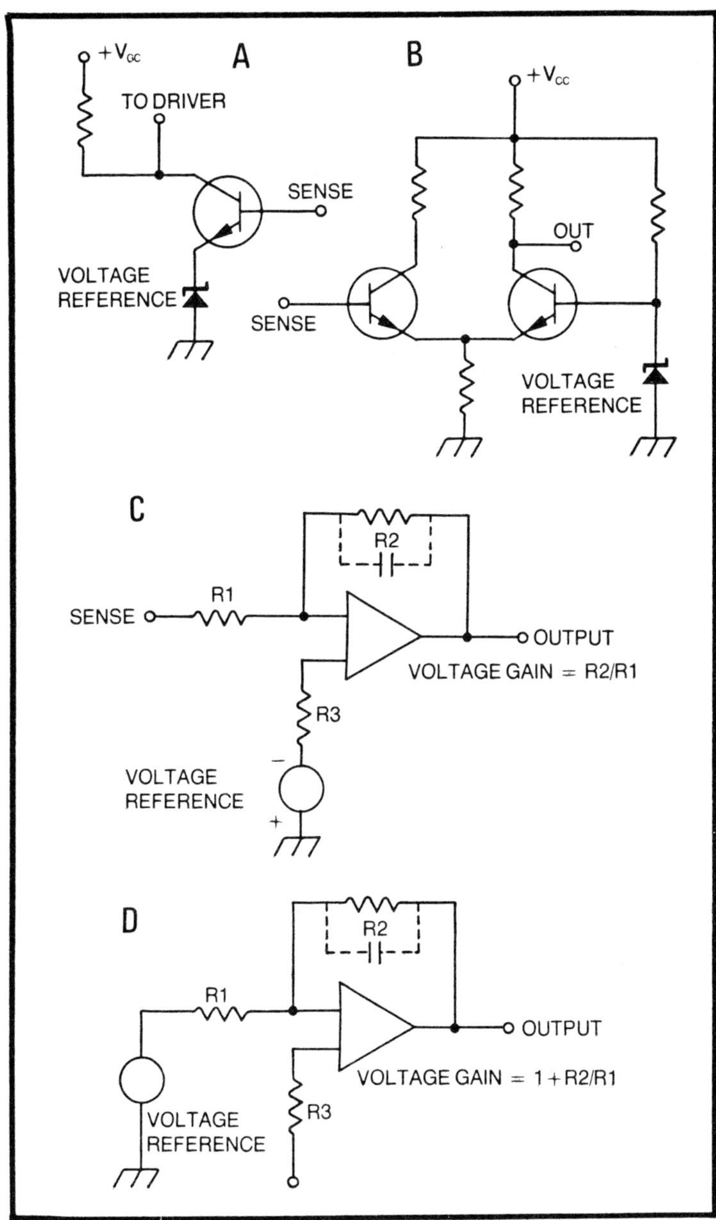

Fig. 9-9. Simplified circuits of nonsaturating voltage comparators. In op-amp circuits C and D, the capacitor placed across resistor R_2 is used to roll off the high-frequency response, and so insure a stable phase−gain margin. For optimum temperature stability, resistor R_3 should equal the parallel equivalent of resistors R_1 and R_2.

the requirements imposed by high-frequency and high gain. And much evidence has been accumulated that points to improved reliability with respect to the use of discrete devices. Even an inexpensive 741 op-amp can provide excellent results in a great many systems.

Two Operational Modes of Comparators

In linear regulators, the comparator itself exhibits a linear transfer function. Although emphasis has rarely been focused upon linearity, the fact remains that such a comparator is essentially a class-A amplifier with dual inputs, it may be inverting or noninverting, and it may comprise one or more active devices. This type of comparator also finds employment in certain switching-type power supplies. The term "comparator" stems from the fact that the output voltage (or some portion of it) is sensed by one of the inputs and *compared* to a stable reference voltage applied to the other input. But this is not the only way of generating a usable response or employing discrete devices or IC modules.

When one inspects the block diagrams and schematics of self-oscillatory switching supplies, it may not be immediately evident that the comparator is operating in a switching mode. Some comparators are three-state circuits—with no error being sensed, the output is zero; and with an input error in one direction, the output is saturated at one polarity; and the converse prevails when the input signal departs in the other direction from the reference voltage. In actual use, a two-state, *on/off* operation would more accurately describe the circuit events. Regulatory action occurs as the consequence of the relative time spent in the *on* and *off* response states. The switching transistor is controlled by the output of the comparator to correct the average DC output level of the supply. Regulation is then the result of appropriate variations in the duty cycle of the switching process. Despite the different operational methods of comparing, the switching regulator performs in very much the same way. (It should not be thought that the saturating-mode comparator is a flip-flop; it remains a linear amplifier, but one of great sensitivity that is deliberately over-driven.)

144

Considerations in the Selection of the Comparator

Whether the comparator in the switching-type power supply will comprise a single transistor, operational amplifier, or special IC module is a decision that must be deferred until the operational mode of the comparator is determined. In linear regulation schemes, the comparator usually provides amplification over a wide voltage range, and the demands imposed upon it can be quite different from the comparator which responds to an error signal by saturating in one or the other polarity. In the first instance, we deal with a linear error amplifier, and we expect amplified versions of deviations from a reference level. This calls for a device capable of voltage or current gain, with the added stipulation that polarity sense must be conveyed, and that is not a difficult requirement to meet. It turns out that the amplification need not even be truly linear over its dynamic range—as is often the case with a single transistor. Such a sense or error amplifier must *not* be driven into saturation. Its frequency response must be sufficient to handle the response time of the overall regulator; it must be free from self-oscillation and from major contributions to feedback-loop instability These are not formidable obstacles.

It is only natural to consider the operational amplifier as a nonsaturating, linear comparator. By virtue of its differential inputs, high open-loop voltage gain, low-impedance single-ended output, and other compelling features in the realm of cost and packaging convenience, these devices appear tailor-made to the needs of the switcher. Among those which have already proven their reliability in the field are National Semiconductor types, 709, LM101, LM741, and LM747. Some op-amps have *internal* frequency compensation, a feature that often simplifies design and manufacture of equipment. On the other hand, an op-amp with provision for *external* frequency compensation can usually provide wideband performance, and often permits an extra dimension of circuit flexibility. For example, it is sometimes advantageous to manipulate the phase gain characteristics of an op-amp to compensate for deficiencies elsewhere in the overall feedback loop.

Another very useful technique for the linear sensing of the error signal involves the use of IC voltage regulators, such as the National Semiconductor LM723 family. Here we benefit not only from the desirable features of operational amplifiers, but from characteristics and modes of performance relating directly to regulating systems. For example, the LM723 contains its own reference voltage; not only is this desirable from the standpoints of convenience and economy, but one would be hard pressed to duplicate its temperature stability with ordinary zener-diode circuits. Moreover, this internal reference has sufficient current capability to provide a stable voltage level for external circuits, with negligible degradation of stability or change in absolute value. This and similar IC voltage regulators have provisions for overload protection, remote control, and programing. A maximum, rather than a bare minimum, of terminals is brought out to allow flexibility in circuit design. For example, the input op-amp comparator may be used in either the inverting or noninverting mode.

The saturating-type comparator operates as a *switch*. The operation is usually open loop in order to advantageously use the tremendous gain attainable. Operational amplifiers have successfully fullfilled this function, but there have been certain obstacles to trap the unwary. At the very outset, the op-amp used in this way must be able to perform without latching up. A latchup mechanism prevents a return from the saturated state. This is not as likely to be encountered as it once was; the problem has been solved in new designs. The response (slew rate) of the saturating comparator should be very fast—even faster than might be deemed necessary for driving the subsequent switching circuits. Fast response is not, as with a power transistor, required to hold down dissipation, but rather to prevent self-oscillation. Recall that the gain is "wide open" between the extremities of bipolar saturation. Often this problem will tax the ingenuity of the designer. A little forethought is generally the best remedy—one must simply be mindful of the routing of the input and output leads from the comparator; another reason why the designer should know not only DC and AC techniques, but also the art of high-frequency layout.

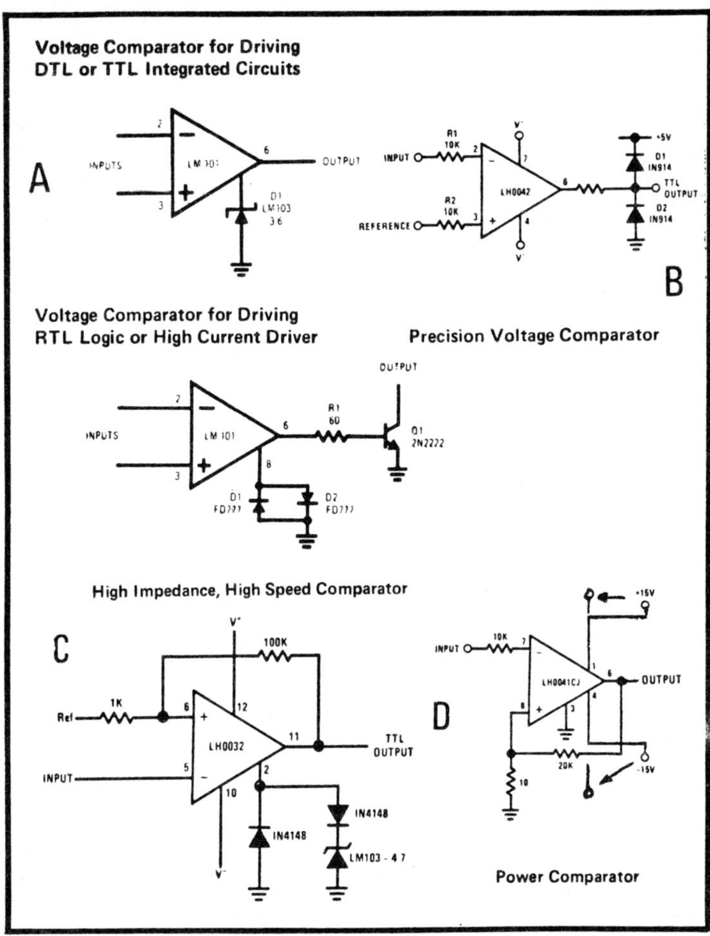

Fig. 9-10. Saturating-type voltage comparators using operational amplifiers. (A) "Workhorse" voltage-comparator circuits. (B) Comparator with FET op-amp. (C) FRT comparator with input-voltage hysteresis. (D) Comparator with high current-drive capability.

Some typical saturating-type comparators using op-amps are shown in Fig. 9-10. In order to enhance stability and increase noise immunity, the saturating comparator often employs hysteresis, so that its operation is somewhat like that of a Schmitt trigger. This is implemented by means of a positive feedback path, as may be seen in examples C and D. Such a technique decreases the sensitivity of the comparator, but is usually found to constitute a desirable tradeoff in switching-type power supplies. The hysteresis characteristic actually provides one of the basic design parameters of such supplies, the peak-to-peak ripple voltage.

A wide selection of op-amps is available for use as a saturating comparator, and many circuit techniques can be applied to speed up, stabilize, or otherwise manipulate the performance. But, the use of specially designed modules should merit serious attention. It happens that much attention was given to the saturating comparator by the semiconductor processors in order to comply with the demands of the computer industry. The results of this concentrated effort are now available in the form of specialized ICs which are optimally suited for use as saturating comparators—even more so than high-quality op-amps. Some of this work has been pioneered by the National Semiconductor Corporation, and has resulted in many excellent IC comparators. Particularly suited to the needs of switching-type power supplies are the LM710 and LM711 families. The fact that these units are compatible with most logic levels is a significant factor in the light of ever-increasing use of gating and logic techniques in the regulating loop of switching-type supplies. The circuit of a monolithic voltage comparator is shown in Fig. 9-11.

A particularly useful comparator is the LM2901, a quad op-amp that operates from a single power supply. Not only are the four individual units useful as saturating comparators, but if only one is required, the others may be connected to provide other functional blocks in the switching-type power supply, such as a one-shot multivibrator, zero-crossing detector, square-wave generator (free-running multivibrator), binary stage (bistable multivibrator), OR gate, and AND gate.

CIRCUIT BUILDING BLOCKS

The versatility of the monolithic IC in general, and of the operational amplifier in particular, is virtually boundless. These devices facilitate the speedy and economical construction of efficient and cleanly performing switching regulators. Depicted in Figs. 9-12, 13, and 14 is a selection of circuits particularly suitable to the many needs of switchers. A number of different ICs are also shown; some types are

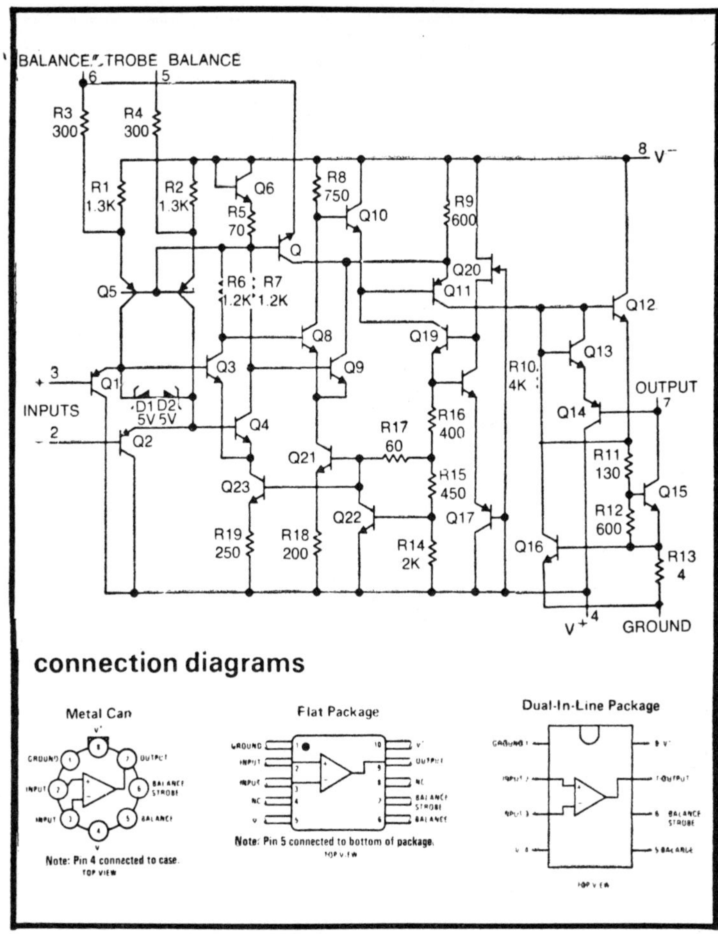

Fig. 9-11. Internal circuitry of the LM311 voltage comparator. The complexities residing in a monolithic circuit such as this quickly rule out the possibility of competing either cost-wise or in performance with discrete elements. Indeed, it is preferable to use such specially designed comparators rather than conventional op-amps when better-than-average results are sought in the performance of the switching regulator. (Courtesy National Semiconductor Corp.)

interchangeable, but each type is endowed with its unique operational features.

The circuitries of Fig. 9-12 are linear circuits yielding proportionate response to input signals. In contrast, those of Fig. 9-13 saturate during their operational cycles, and thereby produce pulse-like waveforms. The group of circuits in Fig.

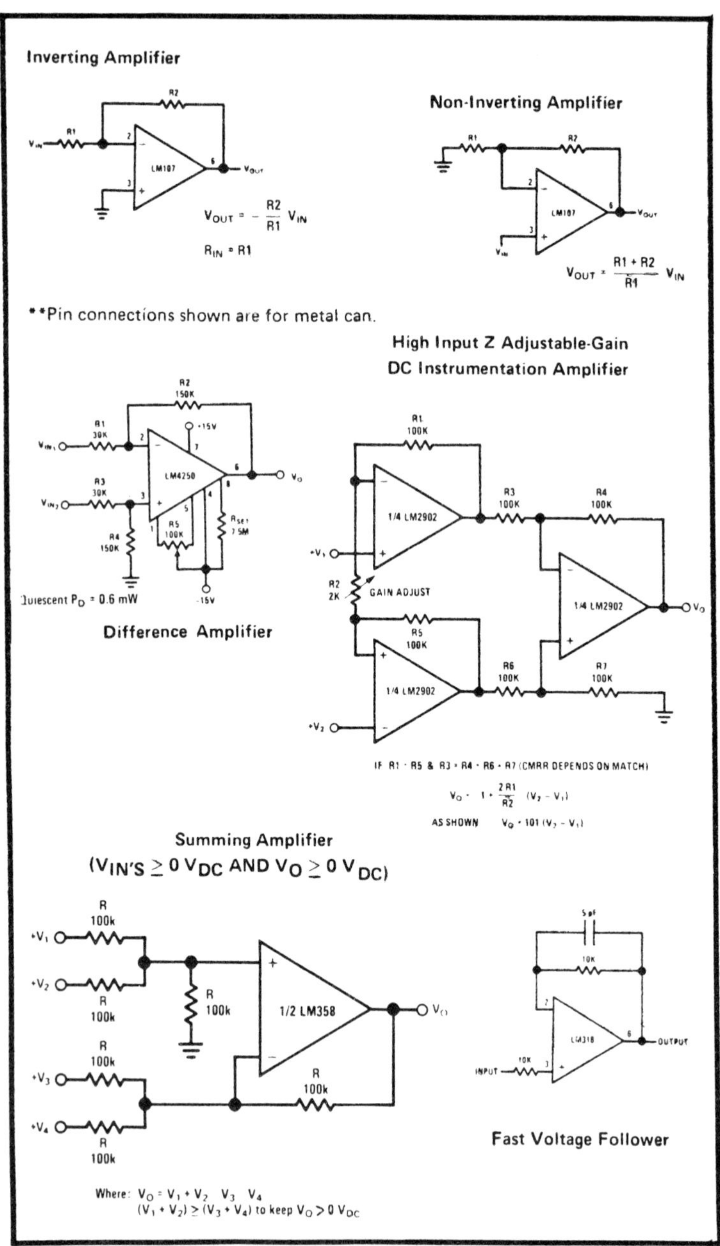

Fig. 9-12. Useful linear circuits configured around operational amplifiers. These circuit functions will be found in many applications involving the sensing, comparing, and amplification of the error signal. (Courtesy National Semiconductor Corp.)

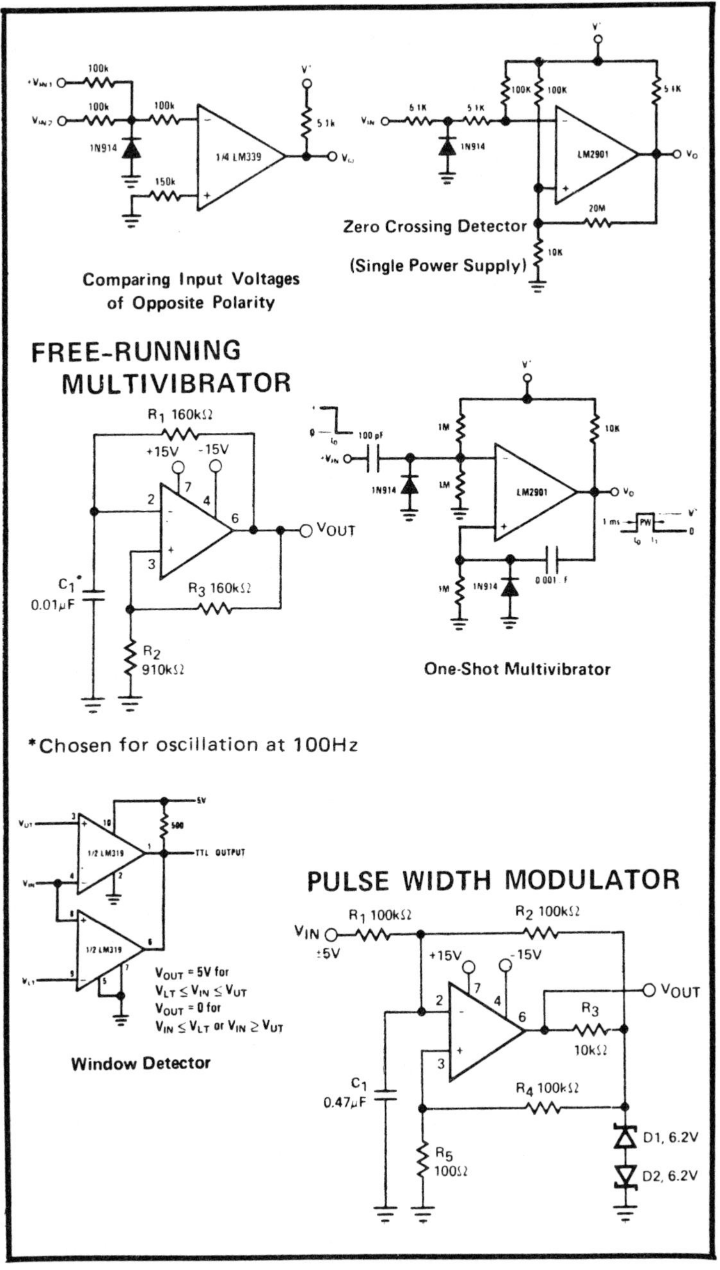

Fig. 9-13. Useful circuits involving saturable operation. These and numerous derivatives are commonly encountered building blocks of the switching-type power supply. (Courtesy National Semiconductor Corp.)

9-14 are combinations presenting six circuit functions that are potentially applicable to various tasks in the control section and feedback loop of switching-type power supplies. Many of these ICs were originally developed by the National Semiconductor Corporation, and numerous applications have been developed by this firm for switching-type supplies. It is suggested that such manufacturers be contacted for assistance in determining the best devices for use in switching-type supplies, or in any other application. New devices are being introduced at an amazing rate, and even the possibility that some new device may better meet your specific needs makes it well worth looking into.

THE LINEAR REGULATOR

Although replacement of linear regulators by switching types is one of the basic themes of this book, it should not be inferred that linear or dissipative regulation of voltage or current is not useful. Once we get beyond the simplest of switching-type supplies, we invariably find that linear regulators have important involvements in the overall switching *system*. There are domains of application where the linear regulator, despite its low operating efficiency, has compelling advantages over switching types. As has been already pointed out, the linear regulator tends to provide closer regulation, less ripple and noise, greater bandwidth, and in many cases, simpler design and implementation. In this book, interest in the linear-regulating technique stems from several considerations.

The building block of switching-type power supplies, such as those depicted in Figs. 9-12, 13, and 14, obviously require sources of DC operating power. Because of monolithic ICs, it is as easy and economical to provide regulated supplies as it is to provide unregulated supplies. The use of regulated operating power leads to better circuit performance because of isolation provided between circuit sections, more predictable performance inherent from stable voltage (or current) sources, and greater attenuation of power-line disturbances. There are other advantages too, such as circuit-control flexibility and programing options. Discrete devices may also

152

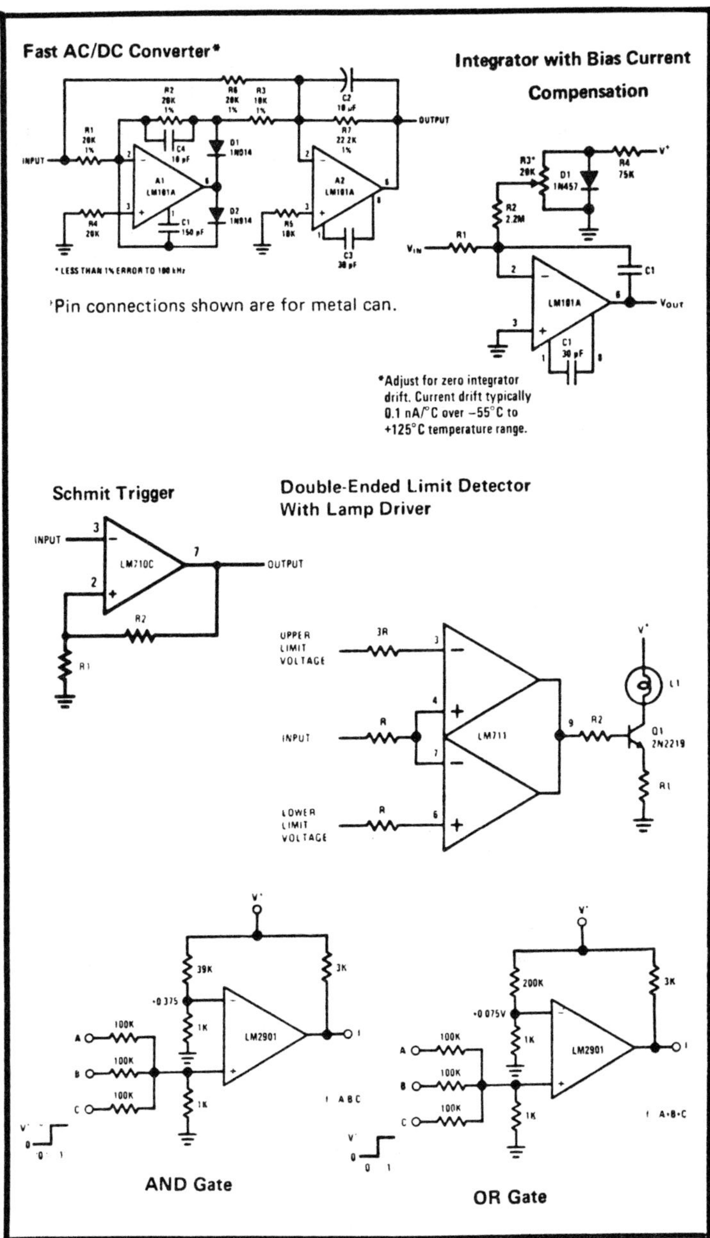

Fig. 9-14. Additional circuits with application potential in switching-type supplies. These and similar configurations are often involved in the duty-cylce-processing portion of the switching-type power supply. (Courtesy National Semiconductor Corp.)

be used, as shown in Fig. 3-5. There is a definite trend, however, to make use of monolithic modules specifically intended for such use, though these are often supplemented by

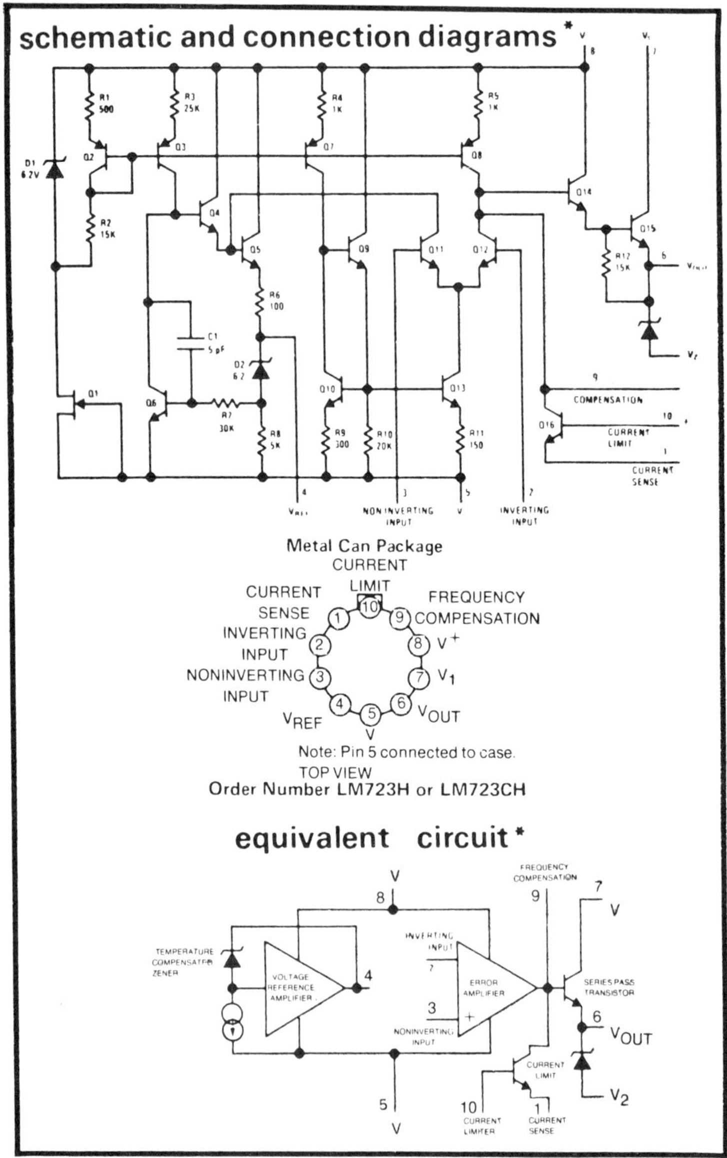

Fig. 9-15. The LM123 linear regulator. (Courtesy National Semiconductor Corp.)

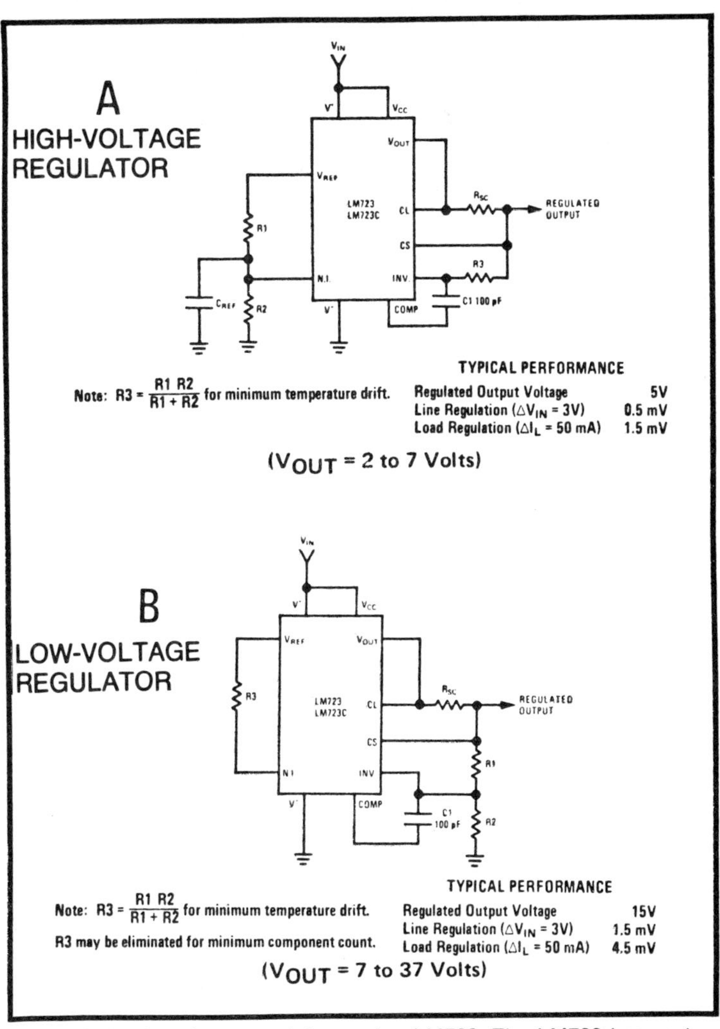

Fig. 9-16. Basic voltage regulators using LM723. The LM723 has an internal 7V reference source, so in A the output voltage is $V_{REF}R_2/(R_1 + R_2)$, whereas in B the output voltage is $V_{REF}(R_1 + R_2)/R_2$. (Courtesy National Semiconductor Corp.)

external discretes which boost the available current or voltage or provide auxiliary control functions. And the use of these ICs has definite advantages in packaging, production, cost, and performance.

The linear regulator appears frequently as part of the overall system of regulation, even though switching

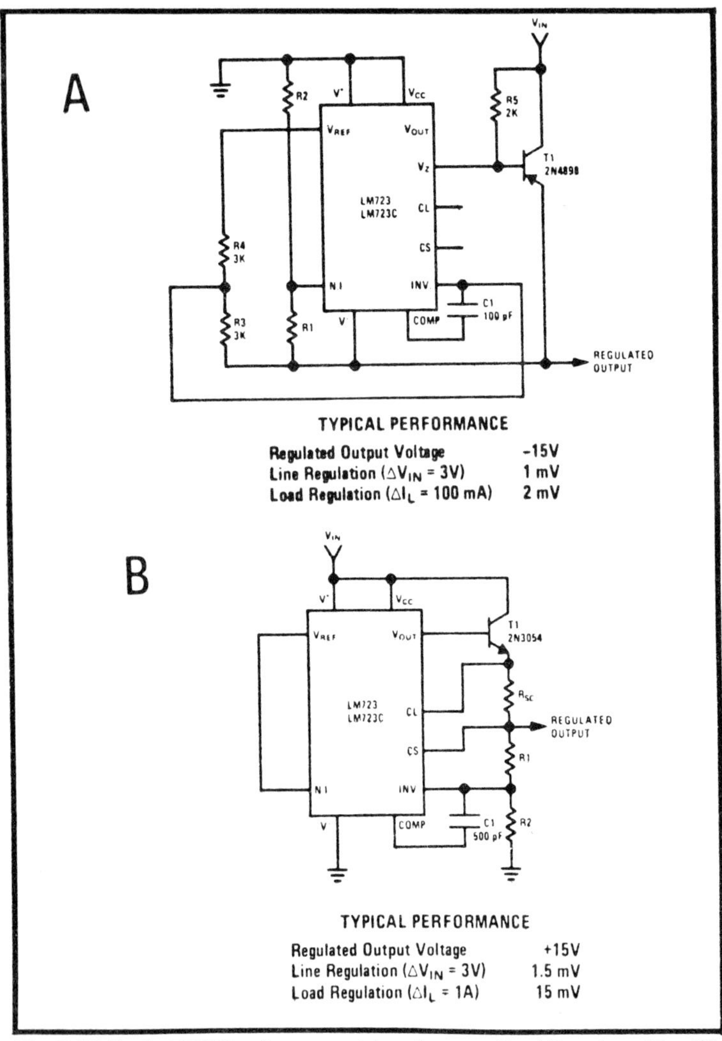

Fig. 9-17. Basic LM723 voltage regulators for negative (A) and positive (B) outputs. (Courtesy National Semiconductor Corp.)

techniques are heavily involved. (The block diagram of Fig. 3-1A shows such an arrangement.) The linear regulator is sometimes advantageously used in the feedback loop of a regulating system.

One of the most versatile of voltage regulator ICs is the LM723 series. This module can provide excellent performance compared to discrete circuits and has few "frozen" design

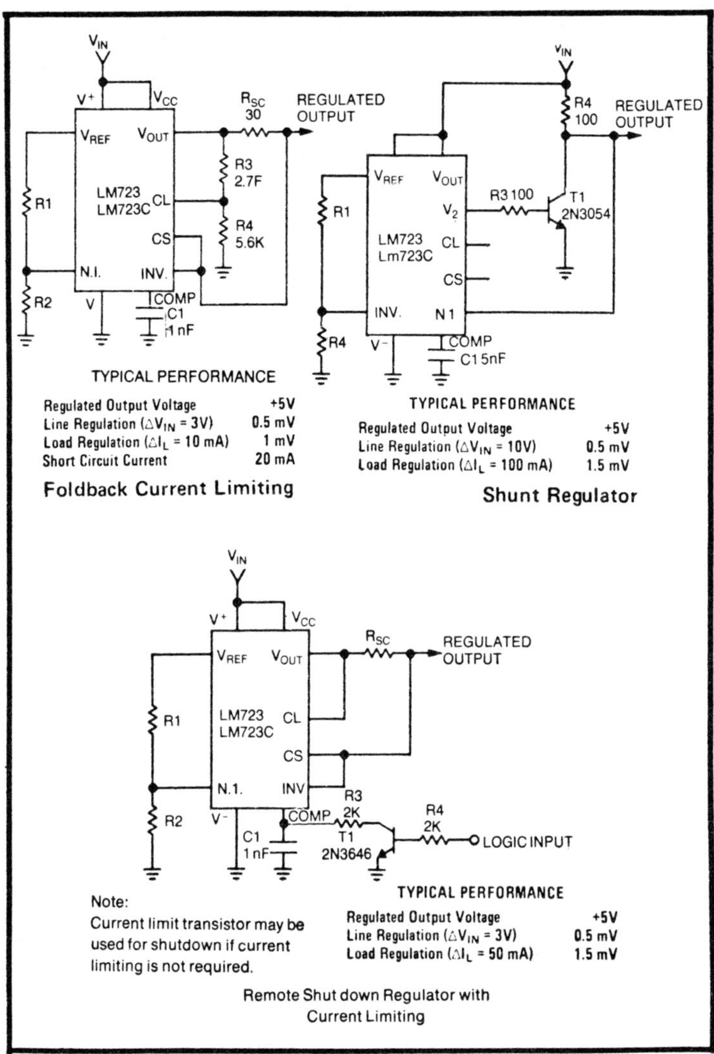

Fig. 9-18. Additional applications of the LM723 linear regulator. (Courtesy National Semiconductor Corp.)

parameters. The manufacturer can illustrate numerous applications and many others can be implemented through the ingenuity of the user. This flexibility is largely achieved by bringing out as many circuitry junctions as is feasible in a standardized package. For example, the voltage reference and the comparator terminals are brought out separately on their own pins. The output impedance of the voltage reference source is very low—enabling its application for other circuit functions with negligible loading effect. The module itself can supply up to 150 mA, and external transistors can easily be added to boost the current capability to many amperes. The circuit is shown in Fig. 9-15, together with connection-pin information.

Basic voltage-regulating circuits configured about the LM723 are shown in Figs. 9-16 and 9-17. The reference voltage is nominally 7.15V with a tolerance of approximately ±5% from unit to unit. Both positive and negative voltage regulators can be readily implemented. Additional applications of the LM723 are depicted in Fig. 9-18. Not only do foldback current limiting, current limiting at a constant value, and remote shutdown constitute protective techniques from the effects of over-voltage, over-current, and short circuits, but these circuit functions lend themselves to various other control objectives occuring in regulating systems. For example, the remote shutdown circuitry could be associated with a large capacitor in order to provide a "soft" startup. The shunt regulator of Fig. 9-18 has the inherent feature of being short-circuit proof. A shorted output circuit simply causes the regulator to shut down, and when the short is removed, the regulator automatically resumes normal operation—a feature not always easy to attain in the more commonly encountered series-pass type of regulator.

10

Designing Switching Regulators

Consider first the simplest of the switchers—the free-running or self-oscillatory type, such as shown in Fig. 10-1. A designer with experience in linear regulators invariably encounters a few stumbling blocks here. For example, the common-sense approach to good filtering with an LC circuit would be to design a low-pass filter with considerable attenuation at the switching frequency. While this would seem to make sense, it is not the proper starting point in the design sequence. A switching regulator must have a certain amount of output ripple for it to work at all. Indeed, the predominant parameter of the output "filter" capacitor is often its equivalent series resistance, rather than its capacitance. It is more essential that the required output ripple be *supported*, than for the LC circuit to function as a good low-pass filter. Of course, there is a conflict of interest—ideally, it would be desirable to make the ripple arbitrarily low, since a low ESR capacitor would reduce capacitor heating and would improve the high-frequency characteristics of the regulator, but such freedom cannot be permitted by the fact that the output ripple voltage is intimately involved with other important design quantities.

FREE-RUNNING SWITCHER DESIGNS

The fact that output ripple voltage is a basic parameter in the operation of the self-oscillating regulator, rather than

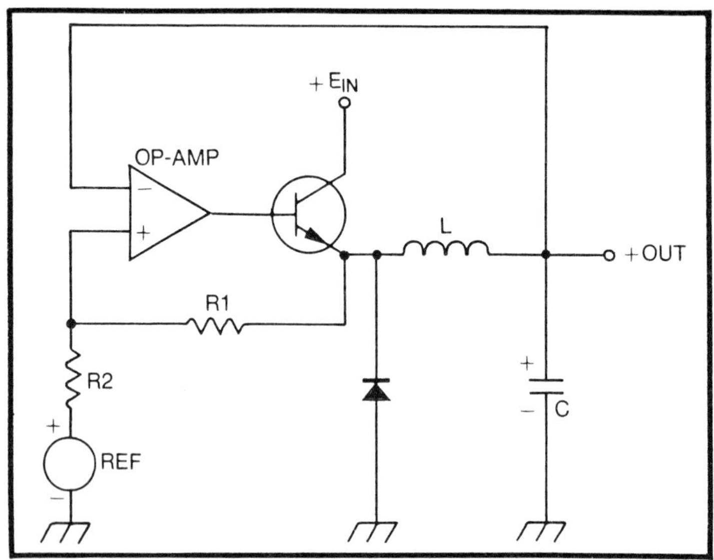

Fig. 10-1. Basic switching regulator. A realistic assumption, especially at high switching frequencies, is that the output ripple voltage is equal to the hysteresis voltage of the op amp.

merely an incidental nuisance to be clobbered with a brute-force filter, can be appreciated from the following relationships:

$$t_{ON} = \sqrt{\frac{(2LC)\,(\Delta V_{OUT})}{V_{IN} - V_{OUT}}}$$

$$t_{OFF} = \sqrt{\frac{(2LC)\,(\Delta V_{OUT})}{V_{OUT}}}$$

$$f = \frac{1}{t_{ON} + t_{OFF}}$$

where f = the switching frequency

t_{ON} = the time the switch is closed

t_{OFF} = the time the switch is open

L = the inductance of the output choke

C = the capacitance of the output capacitor

V_{IN} = the input DC voltage of the regulator

V_{OUT} = the output DC voltage of the regulator

ΔV_{OUT} = the peak-to-peak output ripple voltage of the regulator

160

A useful design equation can be derived from the above relationships, as follows:

$$C = \left[\frac{V_{IN} - V_{OUT}}{(2L)(\Delta V_{OUT})} \right] \left[\frac{V_{OUT}}{f V_{IN}} \right]^2$$

Thus, we have a means of computing the value of the output capacitor in terms of several quantities that are usually known because they are imposed by the conditions of application, or by cost- or size-motiviated choices. The exception here is the inductor L—the value of L should not be determined by resonance formulas. It would not serve any useful purpose to express L in terms of C, since L is determined from other considerations. The waveform diagram of Fig. 10-2 will provide useful insights into the roll played by the inductor.

The inductor must be large enough so that the peak current through it will not be greatly in excess of the maximum steady-state load current. This is to enable the switching transistor to stay within its safe-operating area, and similarly to hold down the peak currents in the free-wheeling diode. At the same time, a moderate peak current relaxes the design requirements on the inductor itself. The basic equation for the inductor is:

$$L = \frac{(V_{OUT})(t_{OFF})}{I_L}$$

where I_L = the peak-to-peak ripple current through the inductor.

also, $I_L = (2 I_0)(n - 1)$ where n is the ratio of the peak inductor current to the maximum average DC output current, I_0.

Using actual numbers, the above procedure will become clear. Assume it has been decided to hold the maximum inductor current to 1.2 times the maximum DC load current. Then, $n = 1.2$, and $n - 1 = 0.2$. For this situation, the equation for L becomes:

$$L = \frac{(V_{OUT})(t_{OFF})}{(2)(0.2)(I_0)} = \frac{(V_{OUT})(t_{OFF})}{(0.4)(I_0)}$$

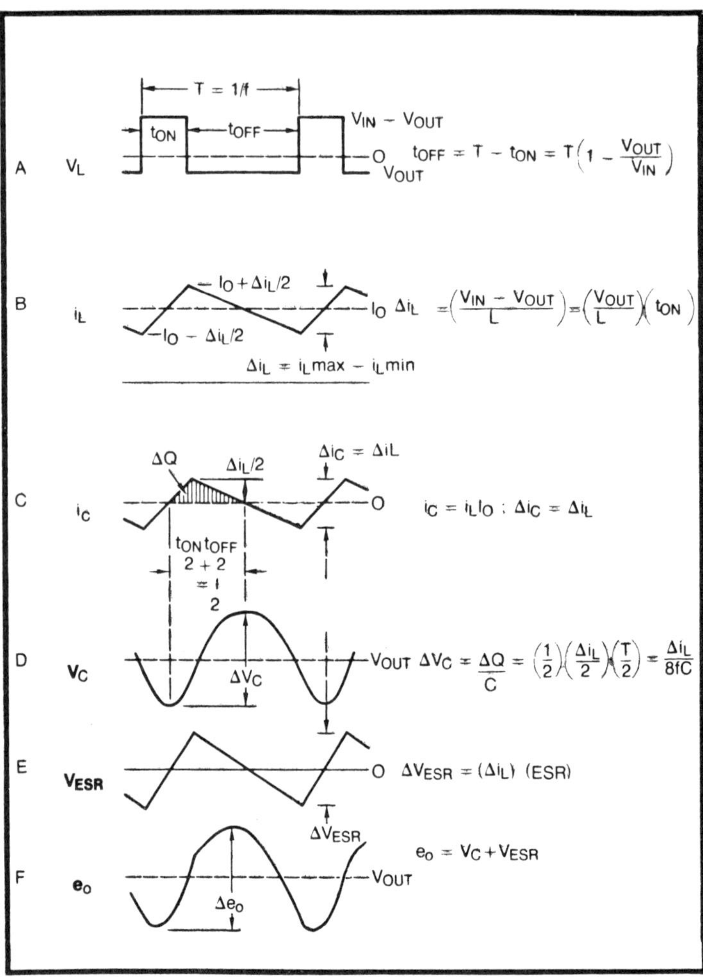

Fig. 10-2. Switching regulator waveforms, showing components of output ripple voltage. In many high-frequency switchers, ΔV_{ESR} predominates over ΔV_C.

The value of t_{OFF} is derived from the relationship,

$$t_{OFF} = \left(\frac{1}{f}\right)\left(1 - \frac{V_{OUT}}{V_{IN}}\right)$$

The parameter that is established from relatively arbitrary considerations is the switching frequuency, f. From the standpoint of reduced size and weight, one chooses as high a switching frequency as possible. But, the overall cost and the speed capability of the switching transistor and the free-wheeling diode act as restraints. The output capacitor also begins to look predominantly inductive at higher frequencies.

It turns out that devices and components with really good high-frequency characteristics are expensive. Yet another frequency consideration is that of audible noise; so, other things being equal, it is often desirable to switch at least at an untrasonic frequency. It has been found that a good balance of the numerous, and generally contradictory factors, occurs in the 20 kHz region. Here, the efficiency is still good with moderately priced elements, and the operation is above the perceptable sound frequencies. In many cases, higher frequencies actually yield diminishing returns with regard to packaging dimensions.

Summing up the design procedure of the free-running regulator, one should first determine the input and output voltage, the switching frequency, and the output ripple voltage. It is to be noted that the output ripple voltage is the consequence of the ripple current in the inductor. Therefore, it is easiest to proceed as follows:

1. Write down the values of V_{IN}, V_{OUT}, and f.
2. Choose a value for n, the ratio of peak inductor current to maximum load current. Nominal values range from 1.1 to 1.4.
3. Comput t_{OFF}.
4. With the information derived in steps *1* and *2*, determine L.
5. Output ripple voltage ΔV_{OUT} is approximately equal to the hysteresis of the comparator. ΔV_{OUT} is also the voltage developed across the capacitor impedance because of the ripple current in the inductor. It is often true that the predominant parameter of the capacitor is not its reactance, but its ESR. If one knows the ESR, the peak-to-peak ripple voltage across the capacitor is approximately the product of the peak-to-peak inductor current and the capacitor ESR. In Fig. 10-2, the hysteresis of the comparator is approximately $(V_{OUT} R_2)/(R_1 + R_2)$.

6. Calculate output capacitor, C. A little contemplation reveals that various LC combinations are possible. High C and low L leads to improved transient response and recovery time; however, too small an inductance leads to high-peak-current demands from the switching transistor at the time the regulator is first turned on, and under certain overload conditions.

THE DRIVEN, SYNCHRONIZED REGULATOR

When a switching regulator is actuated from a fixed-frequency source, its behavior is modified from that of the self-oscillating mode. Most significantly, the output ripple voltage is no longer a basic design parameter; rather, consideration must now be given to the *amplitude* of the drive voltage. (The drive voltage is a triangular wave as it enters the comparator.) The *lower* the drive voltage, the *greater* must be the loop gain of the regulator. But one must not go too far in this direction because noise and transients will ultimately override the error signal and produce erratic operation. Loop instability plagues any feedback system when the gain is too high. And the bandwidth of the regulator tends to decrease with increasing loop gain—producing poor transient response and recovery time. On the other hand, high loop gain does result in better regulation, improved ripple, and faster response to noise transients. The fact that the loop gain is a function of the amplitude of the triangular wave appearing at the comparator leads to a very convenient method of determining loop gain.

It is possible to attenuate the ripple voltage much more in a driven regulator than in a self-oscillating regulator. But, unbriddled enthusiasm is out of place here, because the ESR of the output filter capacitor imposes a limit on capacitor filtering ability. Rejection of input ripple and noise also tends to be inferior to that ordinarily attained from the self-oscillatory mode of operation. The important feature of both operational modes, however, is that inductor L may be considered as an energy reservoir—a source of continuing load current when the switching transistor is off. Accordingly, the inductor may be designed in the same way for both the driven and self-oscillating regulators.

The capacitor, however, need not be calculated with regard to a required output ripple voltage. Rather, it is more convenient and relevant to use the resonance equation,

$$C = \frac{1}{4\,\pi^2 f_C^2 L}$$

in which f_C is a small fraction of the switching frequency, f. Frequency f_C nominally ranges from 1/20 to 1/50 of the switching frequency. Probably a maximum fractional value for f_C would be on the order of $f/5$, but the physically smaller filter components so obtained would constitute a tradeoff for a smaller margin of loop stability. The cutoff frequency of the *LC* filter commonly ranges between 150 Hz and 500 Hz in regulators operating between several kilohertz and several tens of kilohertz. Therefore, the attenuation imparted to 60 Hz, 120 Hz, and 180 Hz input ripple is essentially "electronic" in nature.

GROUNDING TECHNIQUES AND COMPONENT LAYOUT

A common failure in the construction of a switching-type power supply is to assume that because DC and audio frequencies are involved, that parts layout, lead length, and grounding techniques are only of secondary importance. Such an attitude can quickly lead to faulty operation, and it can be faulted for three reasons:

1. Much of the energy exchange occurring in the system occurs in the high-order harmonics. This makes stray parameters important, as well as proximity effects. Thus, one must often approach the matter of physical arrangement in a manner similar to that involved in the successful design of a TV tuner.
2. Very high peak currents can be involved. If these peak currents are coupled into sensitive portions of the circuit, faulty operation almost becomes a certainty.
3. High gain is involved. When a comparator is between its two saturation states, it behaves essentially as a linear amplifier. And unlike most op-amp

applications, the gain of the comparator is "wide open" during its brief transition intervals between saturation, so it is very vulnerable to picking up stray signals at its input terminals. The high loop gain existing in a high-performance regulator also poses a stability problem in itself. It should not be assumed that because there is a positive feedback path in a self-oscillatory regulator, that any further attention need not be paid to the occurance of inadvertant positive feedback—nothing could be farther from the truth. There still remains the error-canceling negative-feedback provision which must meet the stability criteria of any closed-loop system.

Figure 10-3 provides some guidance in parts location and grounding techniques. Of prime importance, the free-wheeling diode and the input bypass capacitor must return together to a *separate* ground, well away from the switching transistor base

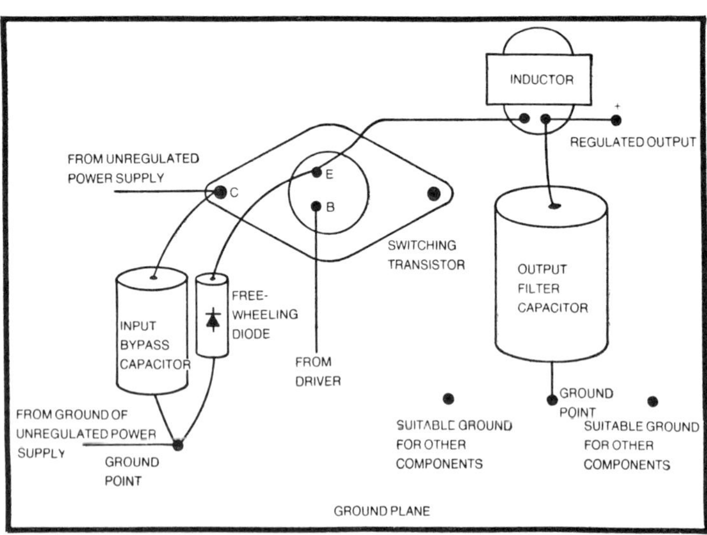

Fig. 10-3. General guidance in parts layout and in location of ground points. In this plan, the high-intensity currents in the free-wheeling diode and in the input bypass (or filter) capacitor do not produce an induction loop embracing the base input lead and other system components, such as the driver, comparator, and reference source. The input bypass capacitor, in some cases, may actually be the output capacitor of the unregulated power supply.

connection. These components should not be grounded at the same ground point used for the output filter capacitor, because if this were done, all space between that ground point (or region) and the unregulated power supply would be embraced by strong circulating current, tending to create inductive pickup in various components and leads. The input bypass capacitor is not always depicted in schematic diagram, and ideally it should not be needed. But in practice, one usually finds that this component is needed, particularly when there is appreciable distance between the unregulated supply and the switching transistor.

VOLTAGE STEPUP IN THE SHUNT SWITCHING CIRCUIT

The shunt-switching circuit of Fig. 10-4 simulates a voltage-stepup transformer. When the switch opens, the energy stored in the magnetic field of the inductor is rapidly released in the form of a large voltage transient. This is the "spike" due to counter-electromotive force, which is advantageously used in ignition, TV flyback, and Geiger-counter power supply circuits. (It is also the source of so-called switching transients in many applications not intended to produce high voltages and the inductance is often of a parasitic nature.) In any event, the voltage transient produced in this circuit is "dumped" into output capacitor C through diode CR. Because of the rectifying action of the diode, the charge accumulated in the capacitor is not discharged when the switch is in its closed position. The capacitor behaves as a reservoir, giving up its energy to the load, but not more rapidly than its rate of charge replenishment.

Not only is the closed switch isolated from the charged capacitor, but it is also isolated from the unregulated DC source, V_{IN}. This is attained by virtue of *limited closure time*. Under such conditions, the current demand from source V_{IN} is in the form of a ramp. Long before the current can become excessive, the switch opens, and a new cycle of operation is ready to start. (Of course, in the event of malfunction, in which the switch remained closed, a short circuit would be produced.) The opening and closing of the switch is under

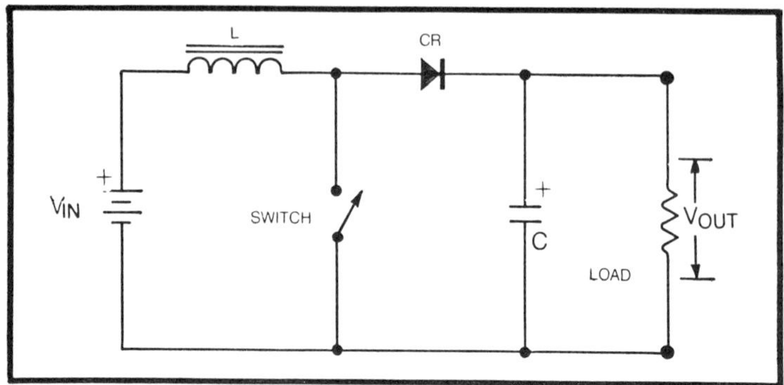

Fig. 10-4. The shunt switcher operates like DC step-up transformer. In this circuit, the output voltage can be considerably greater than the input voltage.

control of a feedback loop and reference. Control is exerted in such a manner that the duty cycle of the switching operation is varied to maintain regulation of output voltage. The simple circuit of Fig. 10-4 serves only to demonstrate the fundamental ideas incorporated in the shunt-switching technique.

The voltage generated by this arrangement is also strongly dependent upon the "Q" of the inductor, and upon the rapidity with which the switch opens. It might appear that such a design might be difficult, but few obstacles stand in the way of successful implementation of this concept. The feedback circuit in actual shunt regulators does not care too much about the *magnitude* of the error voltage—its main concern is with the *direction* of the error voltage.

The design requirements placed on the circuit components in a shunt switcher are not much different than those for a series switcher, and component layout considerations are also much the same. Since the output voltage of the shunt switcher is higher than its input voltage, the switching transistor must be able to withstand this higher voltage. Self-oscillating shunt switchers are not as easily designed as their series-switching counterparts, since the *on* time of the switching element must be limited to prevent excess current through the switch and saturation of the inductor.

If the switch were to remain open, the output voltage would be almost equal to the input voltage, since the voltage drops across the diode and inductor are usually negligible. In addition, the output current under open-switch conditions is also equal to the input current— and none of this current flows through the switching device, which is quite opposite the behavior of the series switcher. Thus, in the shunt switcher, the switching operation is only needed to *supplement* the output power required by the load. (In the series switcher, the switching operation was required to *limit* the power delivered to the load.)

The shunt switcher tends to have lower efficiency than the series switcher, primarily due to the fact that the input currents to the shunt switcher are considerably larger, which contributes greatly to ohmic and switching losses. Nevertheless, in applications requiring power from battery supplies, the shunt switcher is not only able to generate higher power-supply voltage, but it can also *regulate* them—something that a simple DC/DC switching converter cannot do. As a result, shunt switchers are finding many new applications in small battery-operated systems, such as electronic calculators and watches, where small size and efficiency are extremely important.

The basic equation for the inductor in a shunt switcher can be quickly obtained using the equations of the series switcher, by observing the rearrangement of the circuit components in relation to the input and output voltages. Thus,

$$L = \frac{(V_{OUT} - V_{IN})\,(t_{OFF})}{I_L}$$

The *on* time is not usually dependent upon considerations of the LC circuit, but rather upon a maximum or worst-case time limit. Frequently, the *on* time is fixed at a constant value, determined by the equation,

$$t_{ON} = \frac{(L)\,(I_L)}{V_{IN}}$$

and the *off* time is varied to obtain the desired regulation. Such an approach has certain advantages, and it greatly simplifies the startup circuitry of the switcher since the *on* time does not immediately have any effect upon the output voltage. A disadvantage is that the output ripple voltage increases as the average load current decreases, so that the regulation

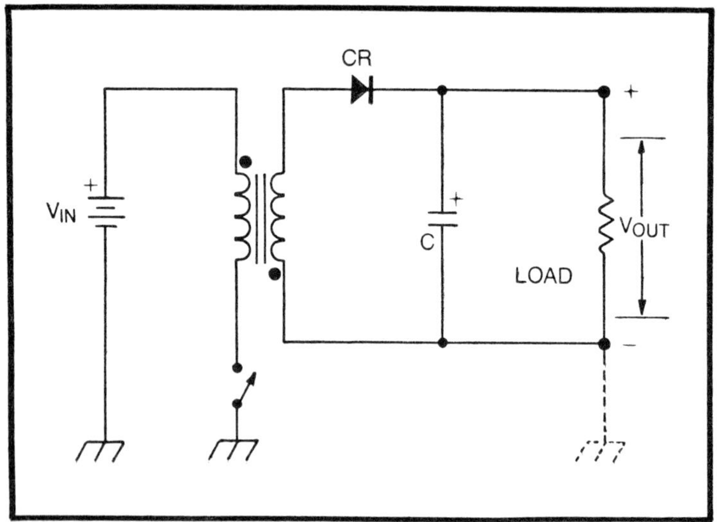

Fig. 10-5. Shunt switching technique for elimination of DC feedthrough. This scheme enables the DC output voltage to be positive or negative, as well as higher or lower than the input voltage.

becomes poorer. An alternative is to choose a lower value for t_{ON}, corresponding to a lighter load condition, or to make provision for reducing or varying the *on* time after startup.

One problem with the shunt switcheris that of DC feedthrough—there is no way of completely turning off the output voltage. In case the switching element fails by opening up, the input voltage passes directly through to the output devices being powered. The shunt-switcher circuit in Fig. 10-4 is also not directly applicable for obtaining output voltages that are less than the input voltage.

A modification of the shunt switcher is depicted in Fig. 10-5, in which the inductor element uses a secondary winding, both for isolation and voltage stepdown. While the output is shown having a negative ground return, it is equally possible to use a positive ground. Note the polarity markings on the inductor windings—the diode must be reverse biased during the *on* time of the switch for proper operation. The design of a good inductor also requires that the two windings be closely coupled magnetically, otherwise large voltage spikes will appear across the switching element, thereby shortening the life expectancy of that device.

170

11

Designing
Rectifier Circuits

Because the switching-type regulating system is much more tolerant of the nominal voltage levels developed by its associated unregulated supply, there are misconceptions in switcher design: that it is only necessary to make certain that there is *sufficient* unregulated voltage under worst-case conditions; that only a primitive unregulated supply with rudimentary filtering is required; and that the switching regulator will take care of "everything else." Admittedly, there is some truth in this, since an ultrasonic switching rate eliminates 60 Hz and 120 Hz ripple in the regulated DC output. The switching process, together with the error-signal canceling action of the feedback loop, provides *electronic filtering*. Life becomes even simpler in some switching systems where the 60 Hz power transformer can be eliminated—a very compelling feature.

If for no other reason, the basic characteristics of unregulated supplies deserve a quick review because they are adjuncts of the switching-type regulating system. But an excessively carefree attitude with regard to unregulated supplies has led to some unfortunate consequences and less than optimum performance.

The important characteristics of rectifier circuits are as shown in Tables 11-1 and 11-2. Considerations of efficiency have made the full-wave circuits the most frequently

Table 11-1. Voltage relationships in rectifier circuits. Factors are normalized with respect to average DC.

Schematic	Voltage Output Waveform	Average DC Volts Output	RMS Volts Output	Peak Volts Output	Peak Reverse Rectifier Voltage	Ripple
Half wave		1	1.57	3.14	3.14	121%
Full wave ct.		1	1.11	1.57	3.14	48%
Full wave bridge		1	1.11	1.57	1.57	48%
3φ Star (wye)		1	1.02	1.21	2.09	18.3%
3φ Bridge		1	1.00	1.05	1.05	4.2%

Table 11-2. Current Relationships In Rectifier Circuits.

		SINGLE PHASE HALF WAVE	SINGLE PHASE CENTER-TAP	SINGLE PHASE BRIDGE	THREE PHASE STAR (WYE)	THREE PHASE BRIDGE
AVERAGE D.C. OUTPUT CURRENT		1.00	1.00	1.00	1.00	1.00
AVERAGE D.C. OUTPUT CURRENT PER RECTIFIER ELEMENT		1.00	0.500	0.500	0.333	0.333
RMS CURRENT PER RECTIFIER ELEMENT	RESISTIVE LOAD	1.57	0.785	0.785	0.587	0.579
	INDUCTIVE LOAD	----	0.707	0.707	0.578	0.578
PEAK CURRENT PER RECTIFIER ELEMENT	RESISTIVE LOAD	3.14	1.57	1.57	1.21	1.05
	INDUCTIVE LOAD	----	1.00	1.00	1.00	1.00
RATIO: PEAK TO AVERAGE CURRENT PER ELEMENT	RESISTIVE LOAD	3.14	3.14	3.14	3.63	3.15
	INDUCTIVE LOAD	----	2.00	2.00	3.00	3.00
		Resistive Load			**Inductive Load or Large Choke Input Filter**	
TRANSFORMER PRIMARY RMS AMPERES PER LEG		1.57	1.00	1.00	0.471	0.816
SECONDARY LINE CURRENT		1.57	0.707	1.00	0.578	0.816

encountered in single-phase power systems. Of the two configurations, the bridge rectifier circuit is usually preferred to the full-wave centertapped circuit. In the first place, the bridge provides a better transformer utilization factor; in the second, the bridge requires no tapped transformer connection. When the bridge rectifier is used in a regulating system that does not use a 60 Hz power transformer, such a system is operable from DC sources as well, and the *polarity* of the DC source is unimportant.

The single-phase bridge rectifier has a few shortcomings, and it may come as a surprise that the centertap arrangement (so popular during the era of tube-dominated techniques) is creeping back into sophisticated switching-type supplies. An increasing demand for regulated DC power at *low* voltage and *high* current is imposed by the computer industry; a common requirement is 5V at several hundred amperes. Neither the conventional silicon junction diode nor the bridge configuration can be readily used for these applications, because the voltage drop they cause greatly lowers rectification efficiency. The bridge is worse than the centertap arrangement in this respect, since the forward voltage drop of two diodes is always involved. The power dissipation in a rectifying diode is proportional to the forward voltage drop, so by merely substituting Schottky diodes for conventional PN junction types in a bridge, a 200% improvement in rectifier dissipation may be obtained. And, if the circuit changes from a bridge to the centertap configuration, there is another 200% improvement in rectifier dissipation. Thus, by dispensing with the commonly used PNP junction-diode bridge, and substituting two Schottky diodes in a centertap circuit, about 400% less power is thrown away in the rectification process. At 30V this may not be of great consequence, but when dealing with only 5V together with high current, the rectifying technique assumes overwhelming significance since the "conventional" approach would in this case totally destroy the high efficiency which a switching supply could otherwise develop.

Although we are dealing with simple circuitry, it is clear that the unregulated supply cannot always be thrown together

174

on the premise that the switching process, feedback, or other techniques will make everything come out well. And, as may be expected, there are other pitfalls. Another problem manifests itself when conventional diodes are used to rectify frequencies above 20 kHz. This has already been discussed in Chapter 8, although the emphasis there was placed upon performance problems rather than just dissipation in the rectifying diodes.

SYNCHRONOUS RECTIFICATION

In the early days of PN junction rectifiers, the germanium diode received great developmental impetus from the major processors of solid-state devices. These diodes, however, became nearly obsolete with the advent of silicon devices. The silicon diode proved far superior in the matters of power handling and the ability to operate at higher temperatures. As if this were not enough, the silicon devices had much less reverse-leakage current and could be made to operate at much higher voltages. However, the germanium junction diode had *one* feature that was superior to the best silicon device: its low forward voltage drop. Designers and experimenters have lamented the demise of the germanium diode whenever confronted with an application requiring *low* forward drop. One such application is the rectification of low voltages at high currents. Fortunately, several of the major semiconductor firms have continued germanium technology, primarily in the form of germanium power transistors. It is well known that these transistors have an inordinately low collector−emitter saturation voltage, $V_{CE(SAT)}$, on the order of 0.3V, even for currents as high as 50A. It would be only natural to speculate on the feasibility of using *transistors* to accomplish rectification.

Motorola Semiconductor Products Company has investigated this possibility and has developed a practical circuit in the form of the synchronous rectifier shown in Fig. 11-1. It has long been known among experimenters that transistors, even defective ones, can be used as a rectifier by using the base−collector junction. But the voltage drop across the collector−emitter terminals of a *saturated* transistor is

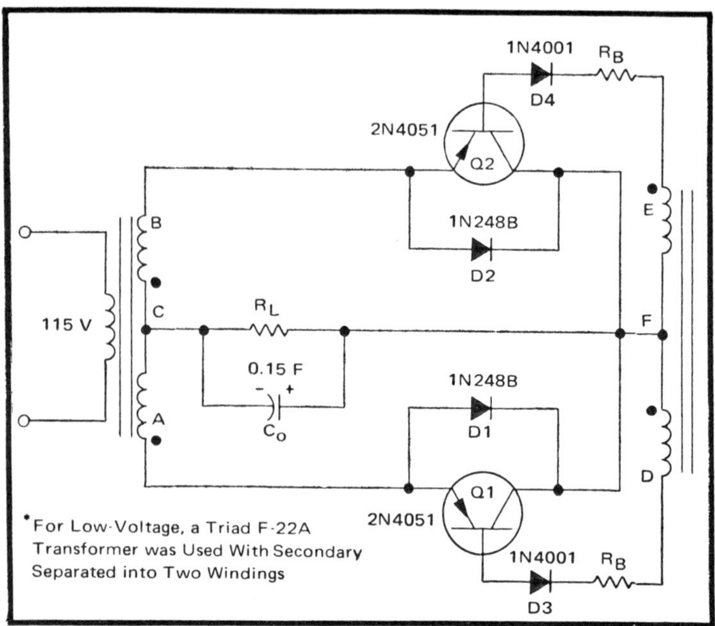

Fig. 11-1. Full-wave synchronous rectifier circuit. The low collector—emitter saturation voltage of germanium power transistors is exploited to produce high rectification efficiency of low voltages at high currents. (Courtesy Motorola Semiconductor Products, Inc.)

even lower than that attainable by using the diode action of the collector—base junction. Thus, it is desirable to use the *transistor action* to make a synchronous rectifier. An interesting comparison of the rectification efficiency of a synchronous rectifier circuit with that of conventional diffused-junction and Schottky diodes is shown in Fig. 11-2. The output voltage as a function of load current for the three rectification schemes is illustrated in Fig. 11-3. Note the exceedingly low output voltages involved. The tests were conducted with a 60 Hz square wave obtained from an inverter.

How much we can increase the frequency and still retain the high rectification efficiency of the synchronous rectifier depends somewhat upon the germanium transistors used, whether a sine or square wave was being rectified, and the nature of the load. It is probable, however, that difficulties from charge storage could be encountered above about 3 kHz. On the other hand, inasmuch as germanium power transistors

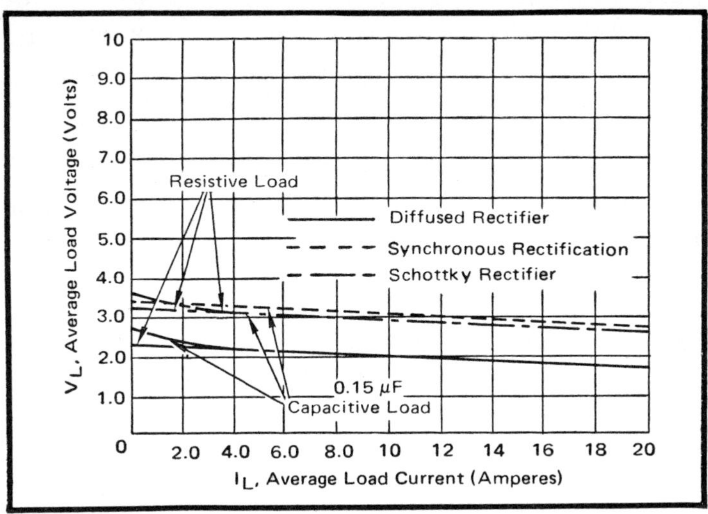

Fig. 11-2. Rectification efficiency of synchronous rectifier. The graph compares the performance of synchronous rectification, with full-wave, centertapped circuits using Schottky and conventional diffused rectifying diodes. A 60 Hz square-wave power source was employed in all cases. (Courtesy Motorola Semiconductor Products, Inc.)

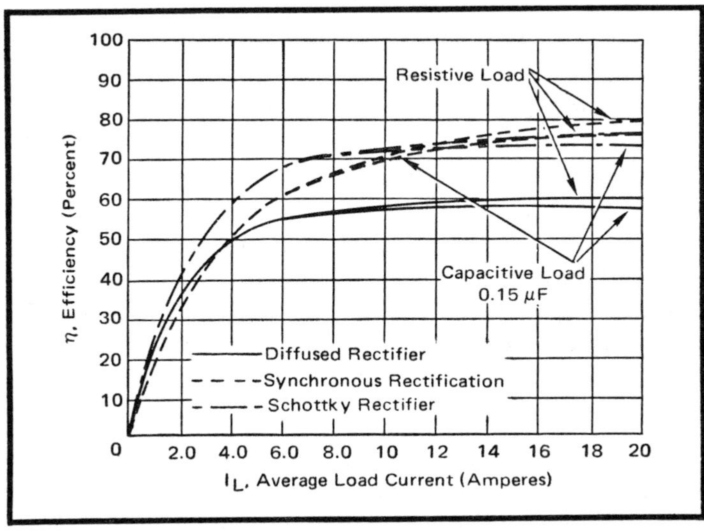

Fig. 11-3. Load voltage versus load current for the three rectification systems. The same centertapped transformer was used for all three full-wave rectification schemes. The synchronous rectifier can excell the Schottky diodes in the matter of low voltage drop, but has a slightly lower efficiency for small load currents. The drop in efficiency is due to the power needed to drive the bases of the germanium transistors. (Courtesy Motorola Semiconductor Products, Inc.)

are used successfully at much higher frequencies in inverters, problems from charge storage could be overcome or satisfactorily reduced using circuitry of the kind used to speed up response in logic gates.

In Fig. 11-1, diodes D_1 and D_2 conduct only when the transistors are forced to momentarily operate in their unsaturated modes, and this operation makes them vulnerable to destruction because the safe operating areas are then exceeded. Such operation usually results from the effect of capacitive loads. Diodes D_4 and D_5 also help reduce this problem by causing earlier turnoff of forward base–emitter bias than would otherwise occur.

An interesting sidelight of the performance of the synchronous rectifier was that the efficiency did not change appreciably when the ambient operating temperature was raised to 80°C. Half-wave versions of the circuit were also investigated. Within the inherent limitations of such rectification, the performance of the synchronous circuit was comparable to that of the Schottky-diode counterpart, and it was considerably superior to that of half-wave rectification using conventional diffused dioees. Thus, synchronous rectification using germanium transistors merits investigation whenever low voltages and high currents are involved. Candidates for such levels of voltage and current are the ECL, IIL, TTL, DTL, RTL, and low-voltage MOS digital systems. In other areas, low voltages or high currents are also required by thermoelectric coolers and often by electromagnetic solenoids. There are many electrochemical processes that fall into this category.

PITFALLS

It is obvious that unregulated power supplies can be quite simple. It has also been pointed out that the series-transistor switching regulator is quite forgiving when it comes to input voltage variations and filtering requirements of the unregulated DC. Yet much grief has resulted from the association of the two units. It happens that the input to the switching transistor looks like a *negative resistance*. It can be mathematically shown that oscillation from positive feedback, shock excitation, and negative resistance are all very similar. Consequently, long leads between the switcher and the unregulated supply can often result in suprisingly strong radio-frequency oscillations. In effect, we have here all of the essentials of a high-frequency, class C *oscillator*—the leads function as a transmission line or tank circuit. The fact that the oscillation is inadvertent does not protect the switching transistor from destruction. Other oscillation frequencies and modes result from the use of LC filters in the unregulated supply. Generally, if inductance is used, a very large filter capacitor must be used to lower the impedance of the tuned circuit seen by the switching transistor. Sometimes it proves rewarding to connect a lower resistance in series with the filter capacitor to lower the Q of the resonant "tank" circuit. Fortunately, it is often possible to dispense with the use of a filter choke; this also tends to be profitable from both a cost and a packaging standpoint.

VOLTAGE MULTIPLIERS

Voltage multipliers have been neglected until recently. Many designers associate these circuits with vacuum-tube technology and thereby tend to overlook some good possibilities. The almost dramatic rescue made by voltage triplers and quadruplers in TV sets is quite well known. Fortunately, we don't have to solve problems relating to X-radiation in switching-type power supplies, but a voltage multiplying circuit can often be useful in further reducing package dimensions after an apparent limit is reached by the more usual techniques of high-frequency switching, elimination of 60 Hz magnetics, etc. In other instances, voltage multipliers can provide a neat method of providing additional output voltage using only a single-secondary transformer.

Many textbooks dwell upon the shortcomings of voltage multipliers. They are claimed to have poor regulation and increased circuit complexities. Statements of this nature involve an element of truth, but are based upon complex vacuum-tube circuits which were always operated from a 60 Hz sine wave. The performance of voltage multipliers is greatly enhanced when driven from square rather than sine

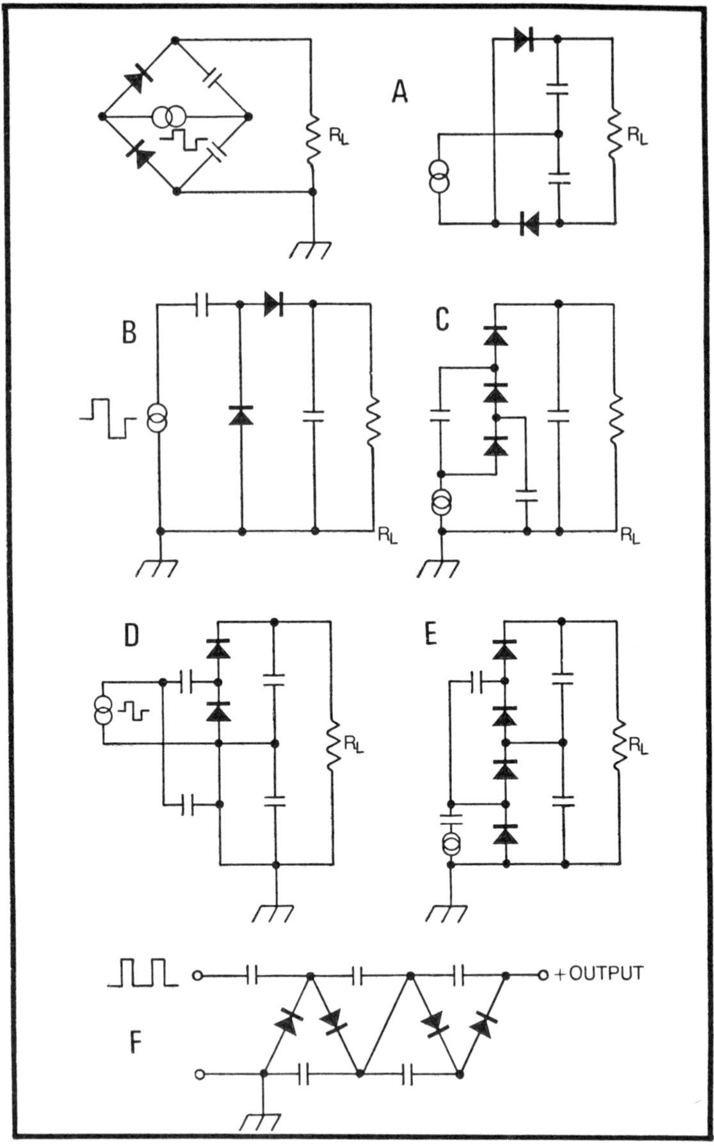

Fig. 11-4. Voltage multiplying circuits.

waves, and especially when driven at higher frequencies. At 1 kHz, and certainly at the popular 20 kHz switching rate, the voltage multiplier deserves reevaluation. Considering that the peak value and RMS value are the same for a square wave, the capacitors in the multiplier circuit have much more time to accumulate charge, contrasted to the peak pulses delivered in sine-wave operation. This manifests itself as improved regulation and filtering. It is true that very good regulation can be had from sine-wave operation too, but only at the expense of larger capacitors. Some useful voltage multiplying circuits are shown in Fig. 11-4. The two topographic versions of the same hookup in (A) show how one can sometimes be misled by the draftsman's technique.

Although poor regulation is no longer a big problem with voltage multipliers, superb regulation is not necessary for use in a system where the ultimate regulation of DC output is taken care of by one or more feedback loops. In particular, the voltage multiplier works very well with 50% duty-cycle inverters. Thus, the voltage multiplier is advocated for use in the unregulated power supply and will normally precede the feedback loop of the regulating circuits. Ordinarily this application is associated with the DC-to-DC converter. For example, the 60 Hz line could be immediately rectified and filtered; this DC could next be used to power a DC-to-DC converter, which in turn could provide the DC input voltage to a switching regulator. Note that this technique enables very high output voltages to be attained without 60 Hz magnetics.

The voltage multiplier also makes it easier to design a good-performing inverter. The inverter transformer works best with a near unity turns ratio. Large deviations from this one-to-one ratio, particularly when stepping up voltage, often produce enough leakage inductance in the transformer windings to cause unstable operation of the inverter. As those who have experimented with inverters and converters well know, oscillation at other than the designed frequency is often the most likely malfunction in what is actually a simple circuit. And the leakage inductance can easily lead to destruction of the switching transistors. This problem can be avoided by the use of a voltage multiplier to permit the use of a near unity turns ratio in the transformer.

When dealing with sine waves, it must be remembered that voltage multipliers operate on the *peak* value of the waves. Thus, a so-called voltage doubler, operating from a

100V RMS input, will show a no-load output voltage of $2 \times 1.41 \times 100 = 282V$. Thus, if the capacitors are large and the load is relatively light, the action more closely approximates tripling of the RMS input voltage. Similar reasoning prevails for the other multipliers.

Assuming equal-valued capacitors and sine-wave operation, voltage multipliers should be designed to have a minimum ωCR product of 100, where ω is 2π times the operating frequency in hertz, C is the capacitance value in farads, and R is the effective resistance in ohms presented by the *heaviest* load which will be imposed. This will result at least in 90% of the attainable DC voltage, and it will confine the operation to a relatively flat portion of the regulation characteristic. For square-wave operation, equivalent results can be had for an ωCR product considerably less than 100.

In selecting a voltage multiplying circuit, due consideration must be given the matter of grounding. In Fig. 11-4, the generator symbol usually represents the secondary of a transformer. Note that grounding of one terminal of the transformer is permissible in half-wave circuits, but not in their full-wave counterparts if one side of the load is to be at DC ground potential. Full-wave circuits are most useful for developing dual-polarity outputs, in which one output is positive with respect to ground and the other output is negative with respect to ground, and half of the total output voltage appears at each output terminal.

12

Designing Inverters and Converters

An *inverter* receives energy from a DC source and produces an AC output. The output is usually, but not necessarily, a square or rectangular wave. (Sine-wave outputs are sometimes obtained by filtering or more sophisticated techniques.) A wide range of frequency and voltage is available by appropriate design procedures. The *converter* is an inverter followed by a rectifier and filter—it is often called a DC-to-DC converter or DC/DC converter. In switching-type power supplies, inverters and converters serve the following functions:

- *Frequency changing*—By transforming the 60 Hz power-line frequency to a much higher frequency, bulky and expensive 60 Hz magnetics can be eliminated.
- *Isolation*—When no 60 Hz transformer is used, these circuits provide isolation between the power line and the DC output.
- *Regulation*—The inverter can be modulated by an error signal, thereby varying its output in such a way as to regulate the ultimate DC output from the system.
- *Voltage translation*—The converter can be used as a DC-to-DC transformer to provide a change in DC voltage level.

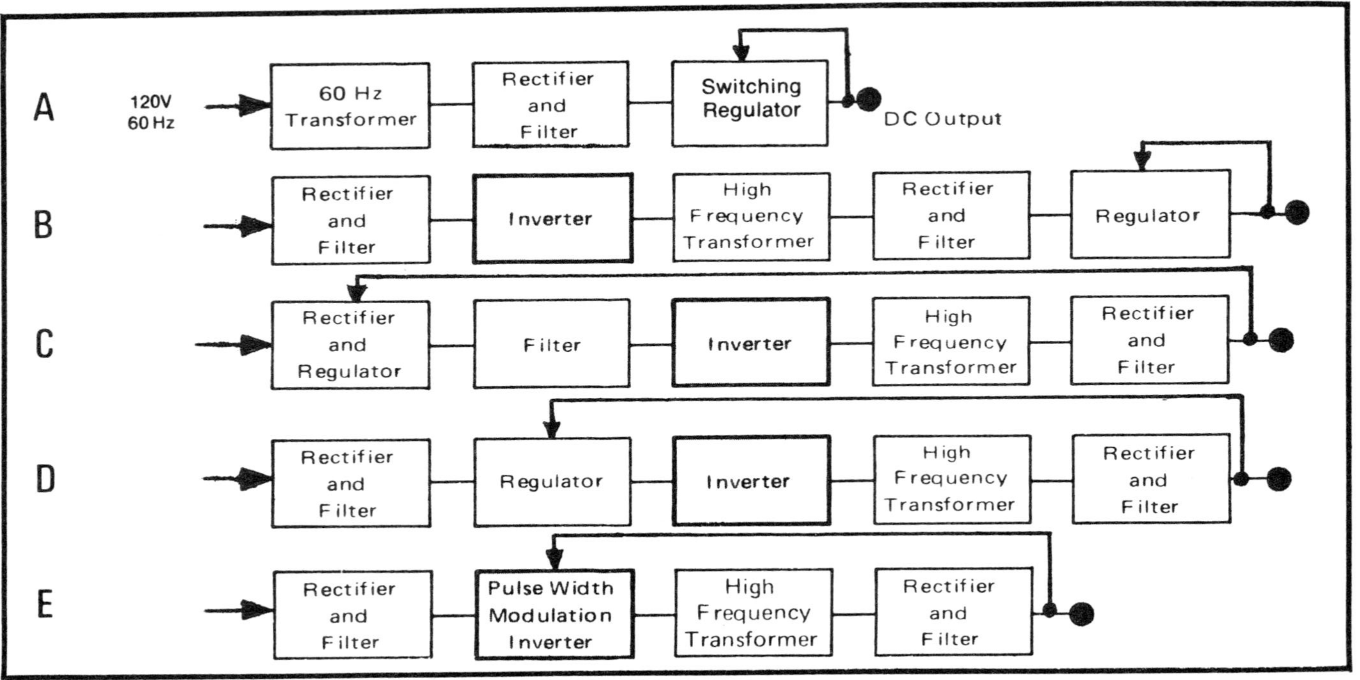

Fig. 12-1. Switching-type regulating systems. System A does not use an inverter. System B, C, D, and E all use inverters. In system E, the inverter is also the regulator.

Figure 12-1 illustrates some of the basic ways in which inverters or converters can be used in power-supply systems. Note particularly the various methods of establishing regulation by applying feedback to different functional blocks.

SATURABLE-CORE INVERTERS

Although there are many variations, the circuits of Fig. 12-2 typify the widely used saturable-core inverter. Sometimes these circuits are referred to as *magnetic* multivibrators; an analogy that conveys the idea that the magnetic core is responsible for the switching action. However, an L/R time-constant does not substitute for the RC time-constant in an actual multivibrator, since such a circuit would not operate with an air core. Not only is a ferromagnetic core needed, but it is desirable that the core exhibit a rectangular hysteresis loop, such as shown in Fig. 12-3. Although the two transistors are ordinarily described (and specified) as *switching* transistors, it is more accurate to state that the actual

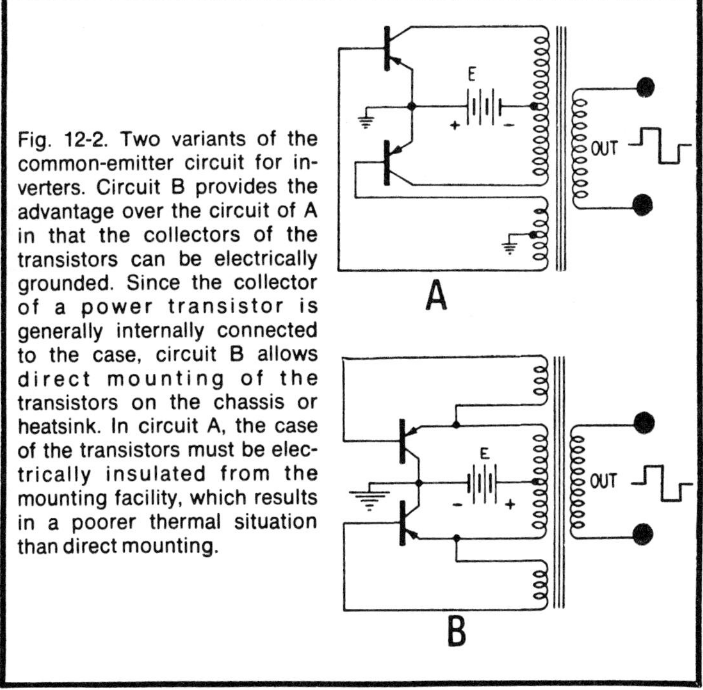

Fig. 12-2. Two variants of the common-emitter circuit for inverters. Circuit B provides the advantage over the circuit of A in that the collectors of the transistors can be electrically grounded. Since the collector of a power transistor is generally internally connected to the case, circuit B allows direct mounting of the transistors on the chassis or heatsink. In circuit A, the case of the transistors must be electrically insulated from the mounting facility, which results in a poorer thermal situation than direct mounting.

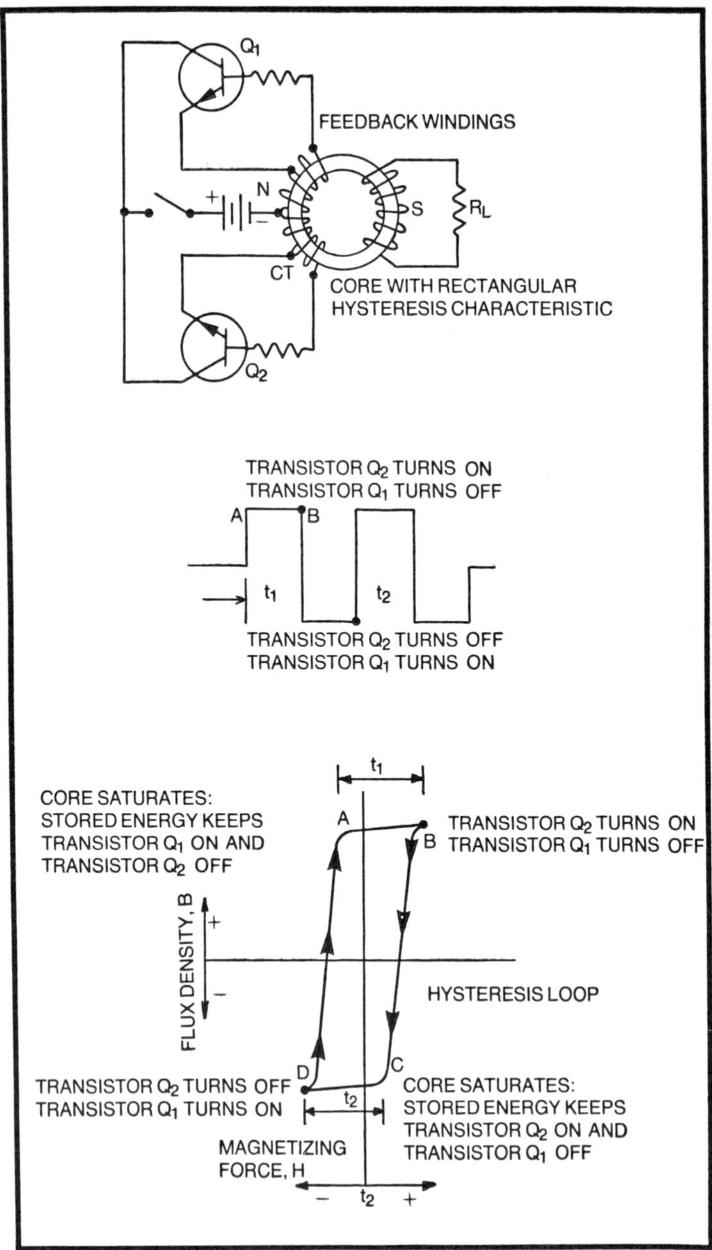

Fig. 12-3. Operational cylce of saturable core oscillator. A large class of inverters and converters used in switching-type supplies operate on the above-illustrated principle. The particular circuit shown is the common-collector configuration.

switching is done by, or at least initiated by, the *core* of the transformer.

When the circuit is first turned on, it is unlikely that both transistors will start to conduct simultaneously, since conduction in one induces turnoff bias in the other. Natural circuit tolerances are generally sufficient to provide for this condition, but it can be reinforced by biasing one transistor differently than the other. If transistor Q_1 initially turns on, then in its *on* state it induces voltage in the feedback winding to maintain transistor Q_2 in its *off* state. That half of the primary winding associated with transistor Q_1 is then subjected to a constant voltage and its flux density rapidly increases. The induced voltage in the feedback windings *reinforce* the conductive states of the two transistors; that is, transistor Q_1 is forward-biased to keep it *on*, and transistor Q_2 is reversed biased to keep it *off*. This situation comes to an abrupt change as the core goes into saturation and electromagnetic induction ceases. An immediate return to original conditions cannot occur because of the energy stored in the core, but the collapse of the magnetic field causes the circuit conditions to abruptly reverse.

The collapse of the magnetic field, now induces voltages to make Q_1 turn off and Q_2 turn on. This action is *regenerative* and gathers momentum until transistor Q_2 is strongly forced into its *on* state. When the core becomes saturated in the opposite direction, the collapsing field again reverses circuit conditions, initiating a new cycle of events. The frequency of oscillation is given by the equation:

$$f = \frac{E \times 10^8}{4\,B_S\,A\,N}$$

where f = frequency in Hz

     E = voltage from DC source

     B_S = flux density at which saturation occurs

     A = cross-sectional area of core (in same dimensional units as used for value of B_S)

     N = number of turns from centertap (one-half of the primary winding)

Typical values for B_S are given in Table 12-1 for a number of different core materials used in inverter and converter circuits.

For designing the transformer, the equation may be expressed in terms of N, to obtain:

$$N = \frac{E \times 10^8}{4 f B_S A}$$

where N is half the number of primary turns.

Strictly speaking, E in both equations should be corrected for $V_{CE(sat)}$, the voltage drop across the transistors, but because of so many variables in the magnetic characteristics of the core, such refinement in computation is not often justified. Higher DC operating voltages tuned to diminish the importance of the transistor voltage drop, although from the standpoint of *efficiency*, transistors with low $V_{CE(sat)}$ should be selected.

The feedback windings are often determined empirically, but typical voltages are in the 3–6V range. A series base resistor is normally used to establish optimum base current. Insufficient base current does not allow the transistor to go deeply into saturation, thereby lowering operating efficiency due to increased $V_{CE(sat)}$. Instability, erratic starting performance, and other ills can result from such "starved" operation of the transistors. Excessive base current again reduces efficiency because of added dissipation in the base–emitter of the transistor, and because of I^2R copper loss in the feedback windings. Another factor to be considered in the design of the feedback windings is the base–emitter

Table 12-1. Saturation Characteristics of Inverter Core Materials.

CORE MATERIAL	SATURATION FLUX DENSITY IN KILOGAUSS (1000 LINES PER SQ CM)
60 Hz Power Transformer Steel	16 – 20
Corosil, Hipersil, Silectron, Tranco	19.6
Deltamax, Orthonol, Permenorm	15.5
Permalloy	13.7
Molypermalloy	8.7
Mumetal	6.6
Ferroxcube 3E2A	½.?

breakdown voltage which can be exceeded by applying too much reverse bias.

The output secondary windings are designed according to ordinary transformer relationships, but it should be kept in mind that the peak-to-peak voltage of the square wave developed across the total primary winding is $2E$ rather than E. The number of turns (N_S) required for the secondary winding is calculated as follows:

$$N_S = \frac{NV_S}{4E}$$

where $N_S = $ number of secondary turns
$V_S = $ peak-to-peak value of the square wave induced in the secondary winding
$N = $ one-half the total primary turns
$E = $ DC operating voltage of the inverter

This relationship is also useful in calculating the number of feedback turns.

Physical proximity between the primary and the feedback winding takes first priority, since leakage inductance here is very detrimental to proper operation. If possible, the entire primary winding should be wound in a single layer. Then the feedback windings should be evenly distributed over those windings to achieve maximum coupling between the windings. Of course, some care should also be taken with the output windings, since good coupling helps minimize voltage spikes on the collectors of the switching transistors and prolongs their life.

PRACTICAL INVERTER CIRCUITS

Some inverter circuits have not changed appreciably in nearly 15 years. On the other hand, some very sophisticated evolution has taken place over those years. The three converters shown in Fig. 12-4 represent applications of the three basic transistor configurations: common emitter, common base, and common collector. The common-collector circuit has the advantage that the transistors can be mounted directly on the chassis or heat sinks without electrical

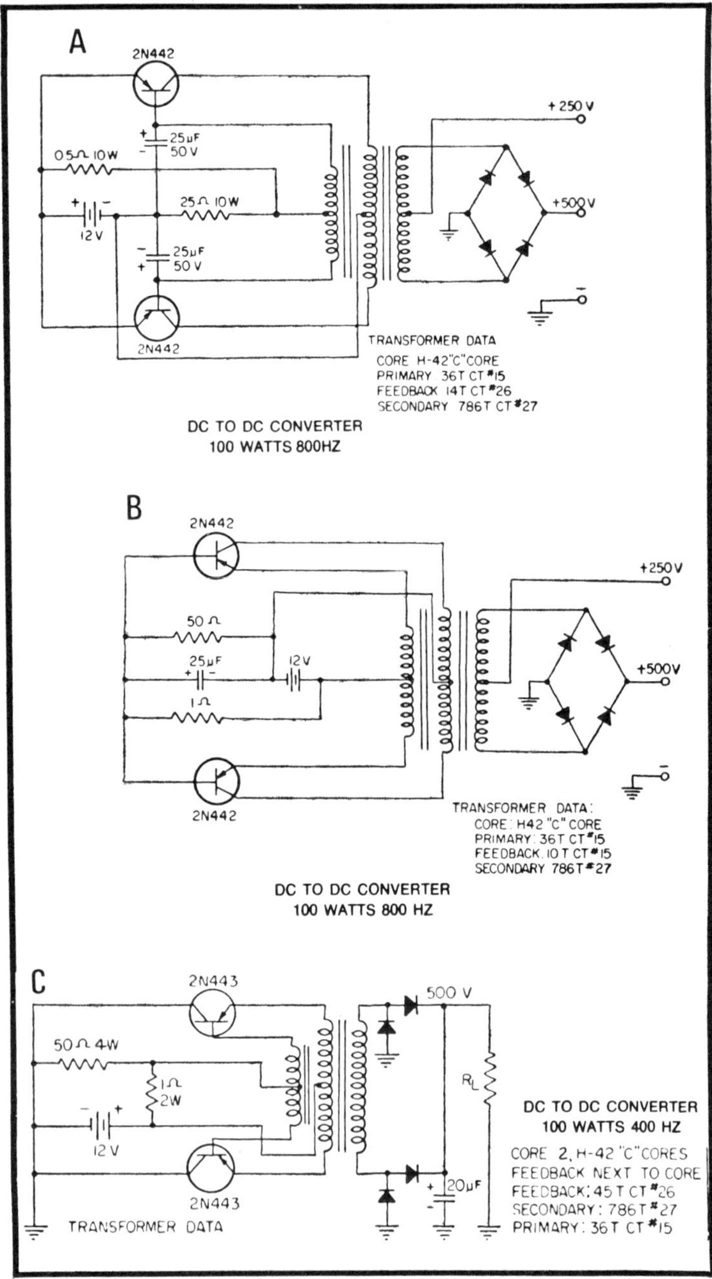

Fig. 12-4. Typical converters using saturable-core oscillators. (A) Common-emitter circuit (B) Common-base circuit (C) Common-collector circuit. (Courtesy Delco Electronics)

insulation. The thermal gradient between the transistor cases or studs and their mounting facility is lower than when such insulation must be inserted. The common-emitter circuit can also be altered to permit such direct mounting as was shown in the simplified common-emitter circuit in Fig. 12-2A. The germanium power transistors used in the circuits of Fig. 12-4 are inexpensive, rugged, and reliable. The "C" cores used in the saturable transformers are easy to wind and assemble. Note the different circuit topography of the output rectifier shown in Fig. 12-4C with respect to those depicted in A and B—all of these rectifier diagrams are *identical*, contrary to the impression gained from first inspection.

The inverter shown in Fig. 12-5 represents a more advanced design, it exploits the superior characteristics of modern components and circuit techniques. Specifically, we see the use of silicon transistors, ferrite toroid magnetics, separate base drive, and oscillation at a relatively high frequency, 25 kHz. In this circuit, the saturation occurs in separate base transformer T_1—transformer T_2 does not saturate. The large ouput available from the small physical dimensions of this inverter well attest to the combined effectiveness of modern components and techniques.

The circuit of Fig. 12-6 is interesting in that the inverter has both its AC output voltage and its oscillating frequency regulated. The ability to control the output of inverters has become an important technique in various switching-type power supplies. In this case, the base currents to switching transistors Q_1 and Q_2 are varied to accomplish regulation of the output voltage. As a by-product of this action, the frequency is also closely stabilized. A sample of the generated voltage is provided by the REF winding on the oscillation transformer. After rectification, this sampled voltage becomes the error signal sensed by comparator stage Q_4 in the linear regulator where control transistor Q_3 varies the base currents of switching transistors Q_1 and Q_2. The nominal frequency of oscillation is 400 Hz. Voltage regulation may be finely adjusted by trimming resistance R_1; that is, this resistance determines whether the regulating characteristic is positive, zero, or negative.

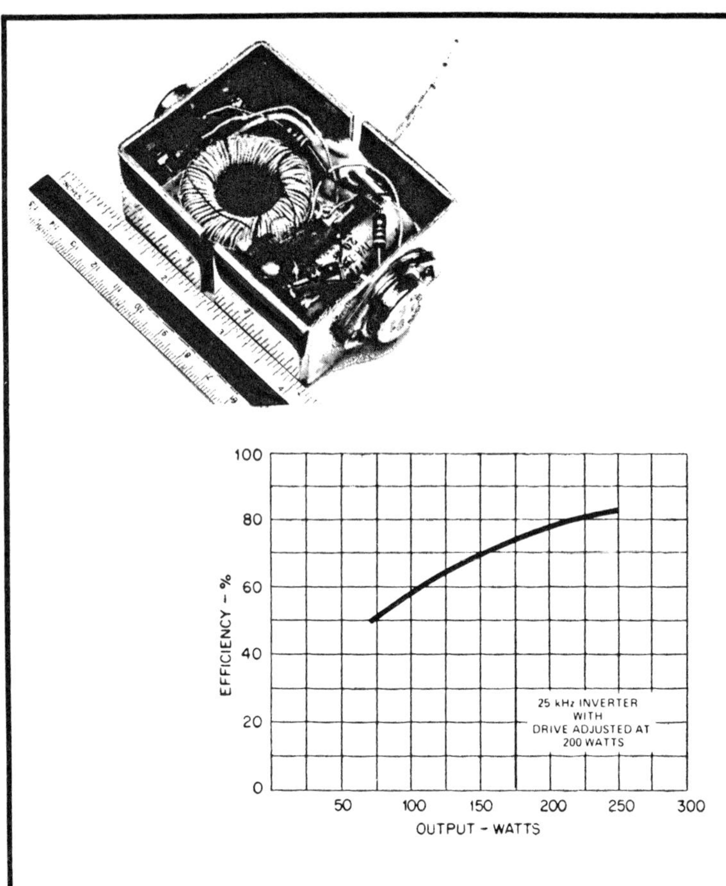

A Line-Operated Inverter

A good example of recently developed switching techniques is illustrated in Fig. 12-7. This circuit constitutes a basic building block for a regulating system that does not use a 60 Hz transformer. Incoming AC power is directly rectified by the bridge, comprising rectifying diode D_1-D_4. The inverter makes use of a saturating transformer (T_1) in the base circuits of switching transistors Q_1 and Q_2. Output transformer T_2 does not saturate. Bilateral trigger diode D_5 insures reliable starting of the inverter. This particular design was specifically intended for powering fluorescent lights—the filtering contributed by capacitor C_1 might be too small for other applications. If this basic scheme is used as the front end of a

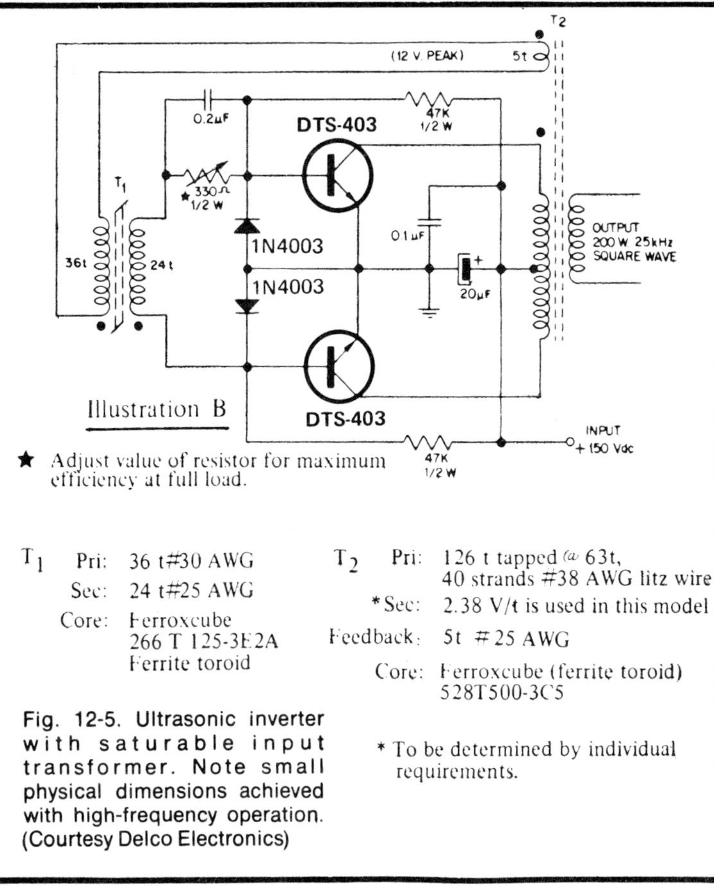

★ Adjust value of resistor for maximum efficiency at full load.

T_1 Pri: 36 t#30 AWG
    Sec: 24 t#25 AWG
    Core: Ferroxcube
    266 T 125-3E2A
    Ferrite toroid

T_2 Pri: 126 t tapped @ 63t,
    40 strands #38 AWG litz wire
    *Sec: 2.38 V/t is used in this model
    Feedback: 5t #25 AWG
    Core: Ferroxcube (ferrite toroid)
    528T500-3C5

Fig. 12-5. Ultrasonic inverter with saturable input transformer. Note small physical dimensions achieved with high-frequency operation. (Courtesy Delco Electronics)

* To be determined by individual requirements.

regulating power supply, it would be advisable to increase the size of C_1, thereby supplying the inverter with smoother DC. About 120W of 15 kHz power is available from winding N_2 on transformer T_2, and as shown, N_2 produces a 120V output. Largely because of the high-frequency ability of the Motorola switching transistors, this inverter attains an operating efficiency of 88% at 100W output. Such performance at line voltage input and at 15 kHz output was in the realm of fantasy just a few years ago.

Line-operated inverters of this type can be fed into a rectifier, filter, and then into a linear regulator. Such a combination can provide an excellent DC-to-DC converter. But for the sake of even greater overall efficiency, the output

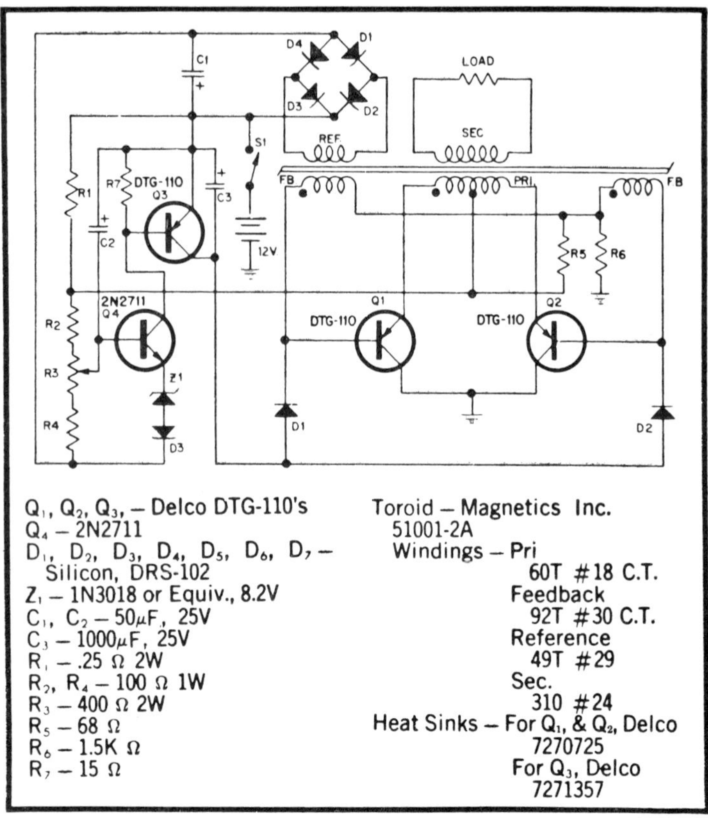

Q₁, Q₂, Q₃, — Delco DTG-110's
Q₄ — 2N2711
D₁, D₂, D₃, D₄, D₅, D₆, D₇ —
  Silicon, DRS-102
Z₁ — 1N3018 or Equiv., 8.2V
C₁, C₂ — 50μF, 25V
C₃ — 1000μF, 25V
R₁ — .25 Ω 2W
R₂, R₄ — 100 Ω 1W
R₃ — 400 Ω 2W
R₅ — 68 Ω
R₆ — 1.5K Ω
R₇ — 15 Ω

Toroid — Magnetics Inc.
  51001-2A
Windings — Pri
    60T #18 C.T.
  Feedback
    92T #30 C.T.
  Reference
    49T #29
  Sec.
    310 #24
Heat Sinks — For Q₁, & Q₂, Delco
  7270725
  For Q₃, Delco
  7271357

Fig. 12-6. Ten-watt regulated inverter. Output voltage and frequency are both regulated. The value of the output voltage is adjusted by means of R_3. (Courtesy Delco Electronics)

regulator can be a switching type configured around a series-switching transistor. In such a case, the switching rate of the output transistor can be synchronized to that of the inverter.

Duty-Cycle Controlled Inverter/Converter

The avant garde of inverter designs is the logic-controlled system shown in the block diagram of Fig. 12-8. The general approach suggested in the block diagram is destined to become of major importance in switching systems. Although the design appears complex compared to simple inverter circuits, the underlying logic is actually quite straightforward, and can be quickly implemented using off-the-shelf digital ICs.

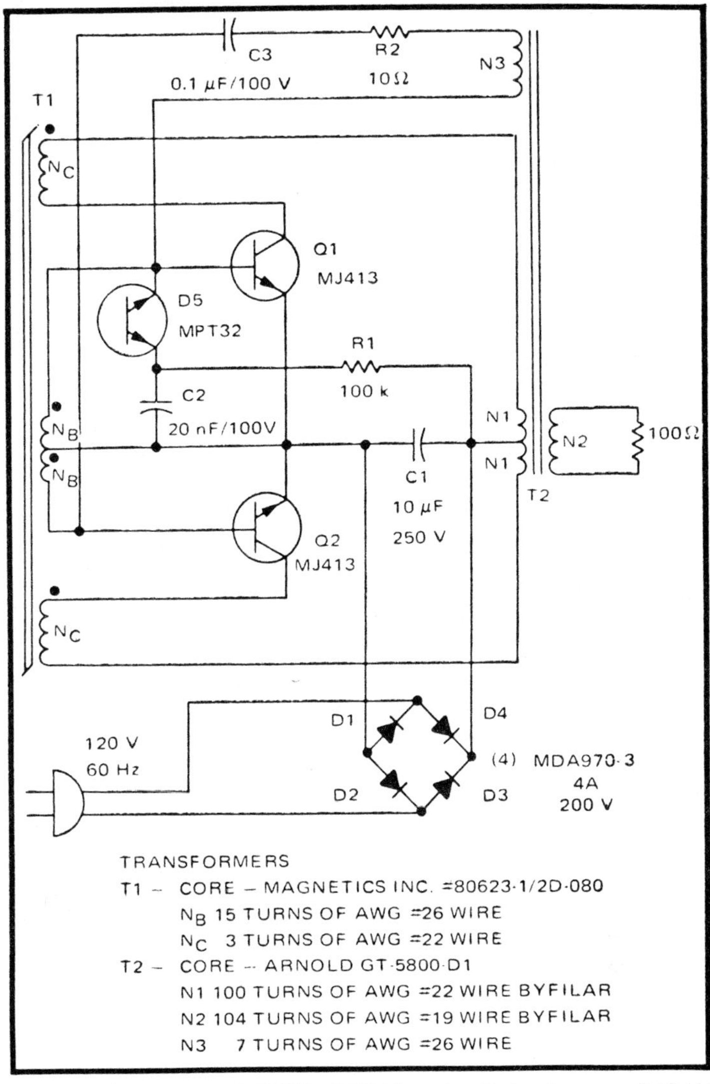

Fig. 12-7. Line-operated 15 kHz, 200W inverter. Note absence of 60 Hz magnetic components. (Courtesy Motorola Semiconductor Products)

To start with, this is a *driven*, rather than a self-oscillatory inverter. The output transformer does not saturate. The driving signal is a square wave obtained from a JK flip-flop, which in turn is triggered from a unijunction-transistor oscillator. A one-shot multivibrator (pulse "on" delay) is connected with NAND gates so that the leading edge of the

195

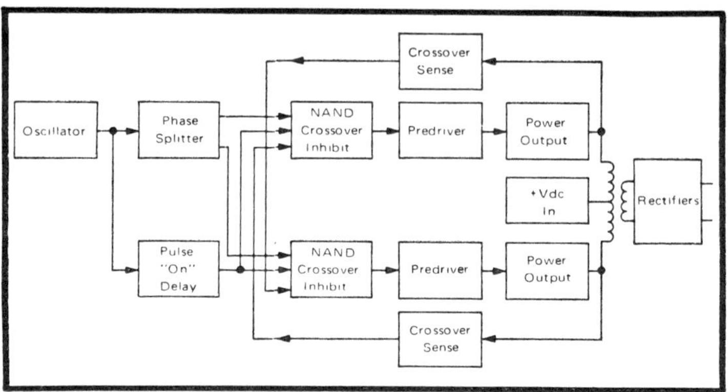

Fig. 12-8. Block diagram of duty-cycle-controlled inverter or converter. A regulating feedback loop is not shown, but a suggested technique is shown in a later diagram. (Courtesy Motorola Semiconductor Products, Inc.)

square-wave driving signals can be delayed. The NAND gates are inserted directly in the path of the drive signal from the JK flip-flop (phase splitter). Thus, the one-shot multivibrator can control the duty cycle of the inverter by either manual or electrical actuation. In the latter case, a feedback loop derived from the output of the inverter could govern the time delay imposed by the one-shot to control overall regulation of the inverter output.

The "crossover sense" functional blocks are peripheral to the basic operating principle outlined above. This does not diminish their importance, however, for they prevent turnon of one switching transistor before the other has turned off. (The simultaneous ON condition of the switching transistors is a commonly encountered failure mode of less sophisticated inverters.) Even when only momentary, this operating mode is undesireable because it tends to exceed the SOA of the transistors. It is a factor that must be considered, largely because turnoff time is usually *greater* than turnon time. Because of the leading-edge delay imposed by the one-shot multivibrator, it might be assumed that the crossover sensing provision is redundant. However, either manual adjustment or electrical control of the one-shot multivibrator could produce an unfavorable duty cycle, such that a conduction overlap could result. The crossover provision protects against this. Note in Fig. 12-8 that this protective provision can override the command of the drive circuitry in the event that one switching transistor has not turned completely off.

From the foregoing discussion, the various items shown in the schematic diagram of Fig. 12-9 should readily fall into place. The 2N2647 unijunction oscillator, together with its phase-splitting circuitry, consisting of a drive transistor and the MC663 JK flip-flop are easily discernable. The variable pulse-width one-shot circuit is comprised of the four NAND gates in the MC668 IC module. The drive-controlling NAND gates are the two sections of the MC671. The output switching transistors are both 2N6308 and these are driven by the 2N6055 transistors. The 2N5088 transistors sense the conduction state of the output transistors and pass this information to pins 5 and 9 of the MC671 NAND gates.

As shown, the output of this inverter is manually controlled by means of the 100K variable resistance in the one-shot multivibrator circuit. However, if the circuit is severed at point X, an electrical input may be injected from a feedback path derived from the output of the converter. Filtering, preferably an LC type, is also needed to further smooth the output. Of course, our inverter then becomes a converter. A suggested feedback arrangement for regulating the converter is shown in Fig. 12-10 in block-diagram form. The MC1723 linear voltage regulator is used in place of a conventional operational amplifier because of its self-contained voltage reference. The M0C1001 optoisolator takes care of such matters as output/input isolation and DC level translation—in essence, this device provides a DC transformer action, but without adversely affecting phase conditions in the feedback loop.

The unijunction transistor oscillates at 40 kHz, but because of the frequency division performed by the JK flip-flop, the output switching transistors are driven at a 20 kHz switching rate. One kilowatt output is obtainable from the circuit. The output transformer used was designed by the Pacific Instrument Corporation of Oakland, California, and provided a nominal 10V output when used with the MBR7230 Schottky rectifiers and an L-section filter (not shown). The timing diagram of Fig. 12-11 provides additional insight into the

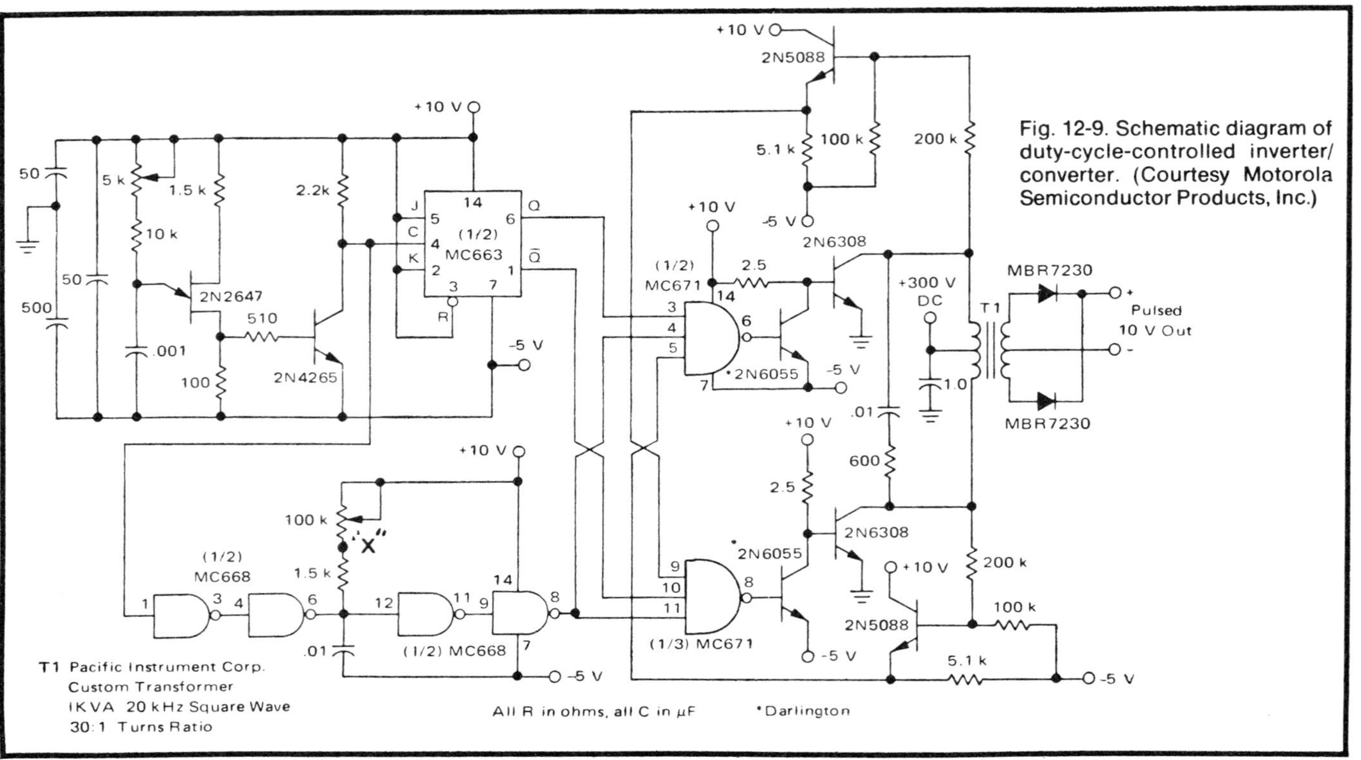

+10 V
2N5088
5.1 k
100 k
200 k
Fig. 12-9. Schematic diagram of duty-cycle-controlled inverter/converter. (Courtesy Motorola Semiconductor Products, Inc.)
+10 V
-5 V
2N6308
50
5 k
1.5 k
2.2k
10 k
50
500
.001
2N2647
510
100
2N4265
J
5
C
4
K
2
14
(1/2)
MC663
3
R
Q
6
Q̄
1
7
-5 V
(1/2)
MC671
2.5
14
3
4
5
6
7
2N6055
-5 V
+300 V
DC
1.0
.01
600
T1
MBR7230
Pulsed
10 V Out
+
-
MBR7230
+10 V
100 k
"X"
1.5 k
(1/2)
MC668
1
3 4
6
12
11 9
14
7
8
.01
(1/2) MC668
-5 V
+10 V
2.5
9
10
11
8
2N6055
(1/3) MC671
-5 V
2N6308
+10 V
2N5088
200 k
100 k
5.1 k
-5 V
T1 Pacific Instrument Corp.
Custom Transformer
IKVA 20 kHz Square Wave
30:1 Turns Ratio
All R in ohms, all C in μF
Darlington

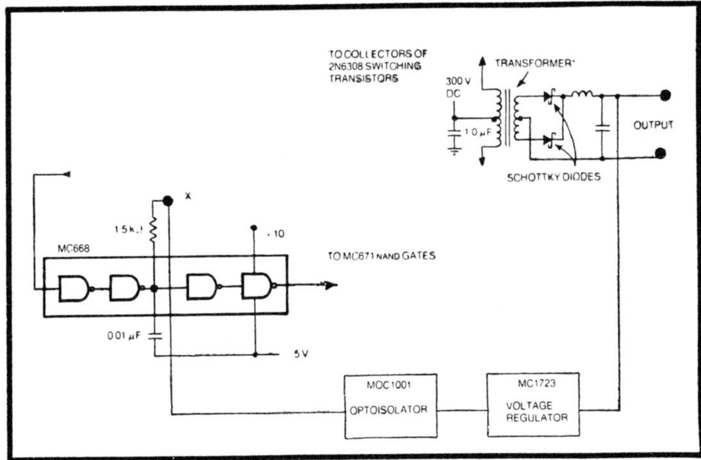

Fig. 12-10. A suggested feedback technique for attaining output voltage regulation. Point X refers to the designated point in the schematic of Fig. 12-13.

unique operational mode of this state-of-the-art system, and should help in relating the schematic circuit to the functions indicated in the block diagram.

SCR INVERTERS AND CONVERTERS

The use of SCRs in inverter and converter applications is attractive because of the high-power-handling capability of these thyristors. A basic problem is *commutation*, for unlike the transistor, the thyristor must be *forcibly* turned off. Unfortunately, this cannot be done by means of the gate; at least not with conventional SCRs, and not at the power levels most generally found in such power supplies. Therefore, the current through the device must be momentarily interrupted. This is usually accomplished by delivering the stored charge in a *commutating* capacitor to the anode—cathode circuit just prior to, or simultaneously with the turnon trigger applied to the alternate SCR. Numerous commutating schemes have been devised, but not all are reliable under all load conditions.

The General Electric Company has pioneered the use of SCRs in inverters and converters. An early circuit is shown in Fig. 12-12. Although improved devices are now available, the basic configuration has served as a prototype for numerous and diversely rated inverters and converters utilizing SCRs.

199

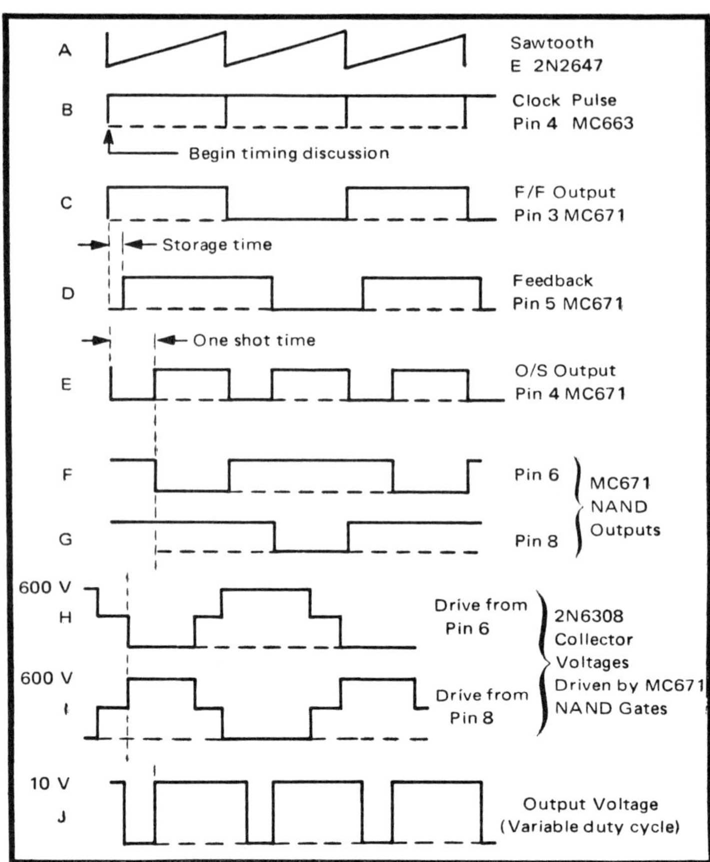

Fig. 12-11. Timing diagram of duty-cycle-controlled inverter/converter. With the addition of a simple LC filter, waveform J can be converted to smooth DC, the level of which depends upon the duty cycle of this wave. (Courtesy Motorola Semiconductor Products, Inc.)

The 1 μF capacitor connected from anode to anode provides the commutation. When one SCR is triggered into its conductive state, a reverse-polarity pulse is delivered by the commutating capacitor to the anode–cathode circuit of the alternate SCR, thereby turning it off. In this type of inverter, the commutating capacitor must be optimized for a certain load, and additional commutating problems arise from inductive loads. But, when used within its load-variation limits, clean operation is obtained and the circuit is both reliable and efficient.

The technical literature dealing with thyristor applications often omits the details of gating circuitry. This is unfortunate,

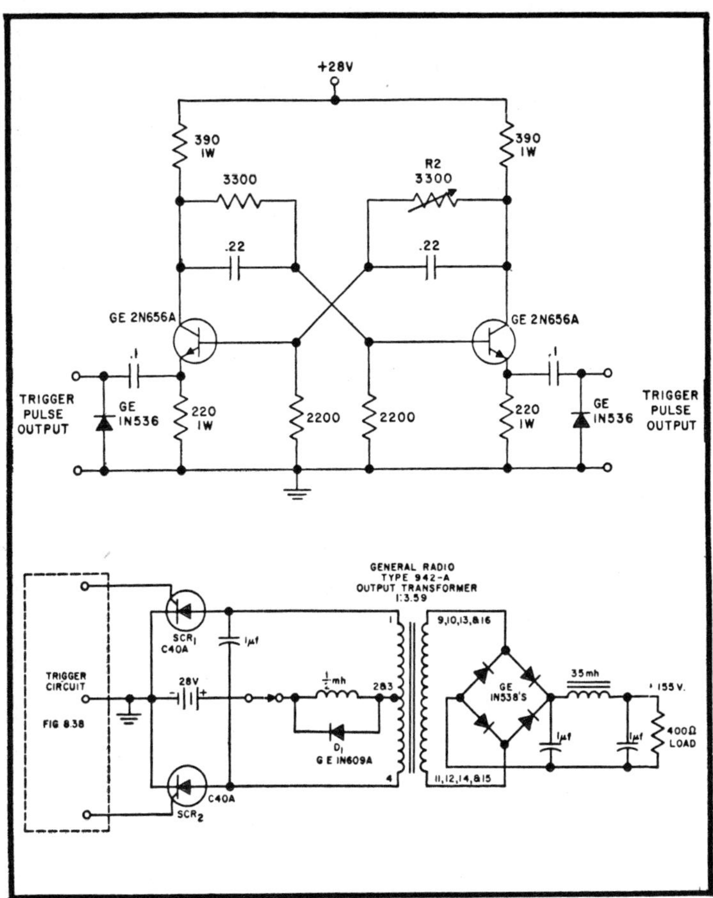

Fig. 12-12. Scr converter and its trigger circuit. This experimental setup paved the way for extensive use of scrs in certain inverter and converter applications. (Courtesy General Electric Semiconductor Products)

for the triggering of the scrs is far from being a trivial matter. In this case, the trigger circuit is shown—a multivibrator arranged to deliver sharply differentiated pulses, displaced 180° in phase.

SCR INVERTER WITH WIDE-RANGE LOAD HANDLING ADAPTABILITY

The inverter circuit depicted in Fig. 12-13 resembles the previous SCR circuit, but incorporates a different commutation technique. In this scheme, the turnon of one SCR shock-excites

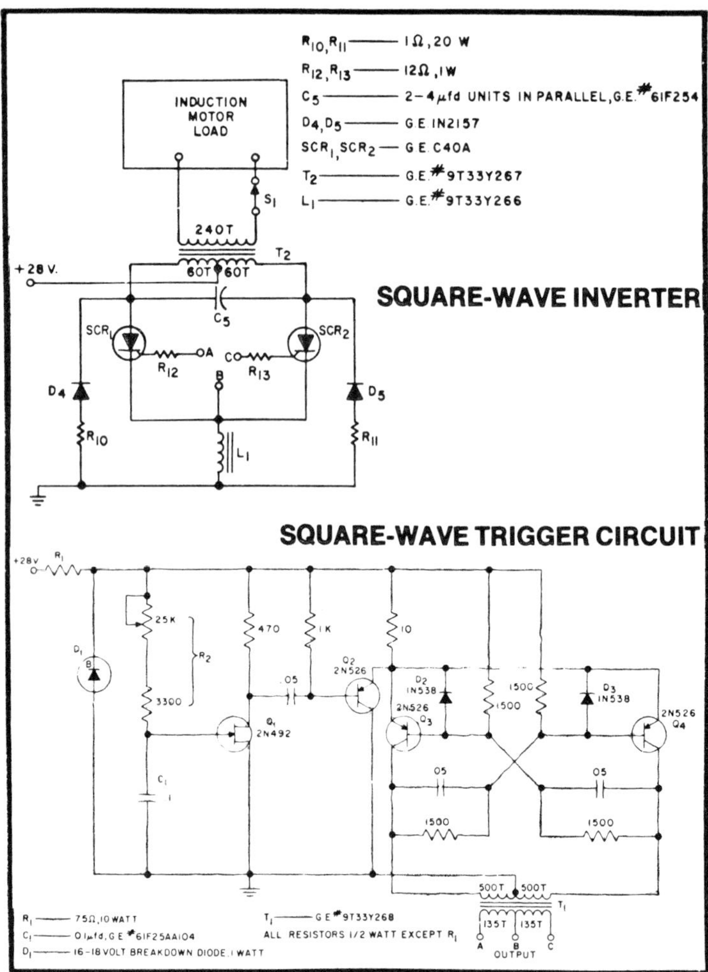

Fig. 12-13. Scr inverter and trigger circuit. This inverter operates well over a wide load range and can be used to power inductive devices, such as motors. (Courtesy General Electric Semiconductor Products)

L_1 and C_5 into series resonance. The oscillatory voltage developed across the alternate scr quickly turns it off. Oscillation is highly damped by resistances R_{10}, R_{11}, and their associated diodes. The diodes maintain the square waveshape of the output waveform by absorbing excess energy. Considerable voltage stress is thus diverted from the scrs, for the peak scr voltage could easily rise to several times the supply voltage without such damping.

What has been previously stated concerning the unique triggering requirements of various thyristor circuits emphatically applies here, for the trigger circuit must be considered as part of the inverter. Merely supplying turnon pulses to this inverter will not enable it to commutate properly when delivering power into inductive loads. What is required is a source of precisely formed square waves. An ordinary free-running multivibrator is too dependent upon its RC timing elements to reliably maintain a 50% duty cycle, particularly when its frequency is varied. The UJT driven multivibrator employed in the square-wave trigger circuit, however, generates complementary square waves from two outputs, each having 50% duty cycles. The UJT pulses are used to synchronize the multivibrator switching transistions. This trigger circuit is also remarkably free of the "jitter" often attending multivibrator operation.

Inductor L_1 tends to be small because a high-Q resonance is not sought in this technique. The use of a small inductor is abbetted by the selection of SCRs with a fast turnoff time. Commutation design procedure first involves the selection of the minimum size capacitor as follows:

$$C = \frac{t \times I}{2E}$$

where C = commutating capacitor, in farads
t = turnoff time of the SCR, expressed in seconds
I = the current handled by the SCR just prior to commutation
E = the DC supply voltage

After the minimum commutation capacitance has been calculated, the actual selection dictates a capacitor somewhat larger—up to several times if permissible by cost and space considerations. Inductor L_1 is chosen to resonate with the commutating capacitor at a frequency corresponding to the half-period which equals or exceeds the turnoff time of the SCR. Although not shown in the diagram, a large capacitor is also desirable across the DC input terminals, but would be unnecessary if it is already incorporated in the DC source and the connecting leads are not too long.

13

Applications of AC Switching-Type Power Supplies

One of the best ways to become familiar with the techniques used in switching-type power supplies is to study the application notes published by the large electronics manufacturers. In order to sell their products, these firms must both develop devices to fit new applications and conceive new applications to optimally utilize devices already developed. Only a well-funded business can finance and maintain such a program. Such firms have already done much of the circuit engineering that otherwise burdens the user of their products. In the specific case of switching-type supplies, it is surprising how much money, time, and effort are needlessly invested by smaller firms in the pursuit of circuit or system objectives already worked out in the superior facilities of the large companies. The knowledgeable worker in electronics can find some interesting and diverse circuits in these engineering notes, and he can adapt portions of them to his own needs or derive techniques somewhat modified from those presented.

GENERAL GUIDELINES

With regard to the self-oscillating type of switching regulator, the following facts are relevant:

- The input noise and ripple riding on the unregulated supply line are well rejected.

- The output ripple voltage may be determined by the designed-in hysteresis of the comparator, but its lower limit is imposed by the ESR of the output filter capacitor.
- The resonant frequency of the LC filter is commonly no greater than about one-twentieth of the switching frequency.
- The ESR of the capacitor is a basic design parameter of the switching circuit. At the same time, it must be kept in mind that a low ESR prevents excessive temperature rise in the capacitor.
- The peak-to-peak current in the inductor must cause neither current discontinuity nor core saturation. A nominal excursion for this triangular wave would be about 10% above and below the DC load current; this is only a ballpark figure, but it is a reasonable one for many applications.

The driven or synchronized switching regulator has, in many respects, a similar circuit configuration to that of the self-oscillating type. Both make use of a transistor switch, an LC output filter, and a free-wheeling diode. There are, however, significant differences in both design and operation:

- Ripple and noise on the unregulated supply line result in poor output regulation, since the unregulated supply adversely affects the DC regulation of the overall system.
- Unlike the self-oscillating mode of operation, driven operation does not impose any requirement for output ripple.
- The ESR of the filter capacitor is not a basic design parameter of the driven switch circuit, but the ESR should be low in order to avoid excessive temperature rise in the capacitor from ripple current I^2R heating.
- In the driven switcher, the loop gain is governed by the amplitude of the triangular-wave drive voltage, which is superimposed upon the reference voltage. Small drive amplitudes correspond to a high loop gain, which is desirable in order to suppress ripple from the

unregulated power supply; however, excessive loop gain makes the comparator vulnerable to other noise sources and can therefore degrade regulation and stability.

- The bandpass response-time characteristic tends to be less than for a similar self-oscillating switcher operating at the same switching rate. This adversely affects response and transient recovery time.

Neither the oscillating nor the driven switching regulator can remove switching frequency ripple by the action of the feedback loop.

Regulators Using Shunt Switching

Conventional shunt switching regulators are characterized by DC feedthrough from the unregulated power supply. This means that the output voltage is available at higher, but not lower, voltage levels than that of the unregulated input voltage. The restriction can be circumvented by using an inductor with a secondary winding. The shunt transistor, in any case, must be able to withstand relatively high peak currents, and this is why the shunt regulator will be encountered only at relatively low power levels. Although there is no hard and fast rule, the shunt regulation scheme has rarely been used beyond the 150W output level; exceptions have occurred where factors other than cost are important. At low power levels, however, useful objectives can be accomplished, such as producing an appreciable voltage stepup. This circuit has recently appeared in electronic calculators, watches, geiger counters, and fluorescent-light power sources. Filtering problems may occur because of the waveshape of the flyback pulses generated when the switch opens. On the other hand, some workers feel that this type of switching supply tends to produce less RFI than does the conventional series switcher.

Inverters and Converters

When working with inverters and converters, the "apparently" superficial capacitors, RC networks, and diodes often associated with the basic circuit should not be ignored. These are generally despiking, snubbing, or energy-absorbing provisions used to protect the active devices from transients, turnon surges, and excessively high rates of change in either voltage or current. Without these circuits, either faulty operation or catastrophic destruction can occur.

Unless stated, or schematically indicated, the output transformer of driven inverters ordinarily does not saturate. In the simple self-oscillatory inverter, the operation is dependent upon transformer saturation, but in more sophisticated types, the requisite saturation occurs in a small base-drive transformer rather than in the large output transformer. Look for the saturating-core symbol on the schematic.

SWITCHER USING LINEAR VOLTAGE REGULATOR

Both the similarities and differences of linear and switching-type regulators have received considerable attention. Indeed, the quick change from the dissipative to the switching mode, depicted in Fig. 3-5, might be considered interesting but not too significant. It happens, however, that the ability to convert from linear to switching-type regulation has been fortuitous in the evolution of high-performance switchers. The National Semiconductor Corporation and other manufacturers of solid-state devices, have developed an extremely useful line of linear voltage regulators; these were specialized adaptations of previously developed operational amplifiers. The linear regulators include internal temperature-stabilized voltage references and external connections enabling such operational modes as current limiting, current foldover, remote shutdown, and drive to high-current booster stages. That these functions are contained in an IC module means that the overall performance greatly exceeds that readily attained by using discrete devices, hybrid modules, or even monolithic op-amps.

In Fig. 13-1A we see the circuit of the LM205. The circuitry of the IC is not meant to duplicate a circuit made with discrete elements. A discrete layout of the indicated components would be attended by formidable conflicts of interest involving

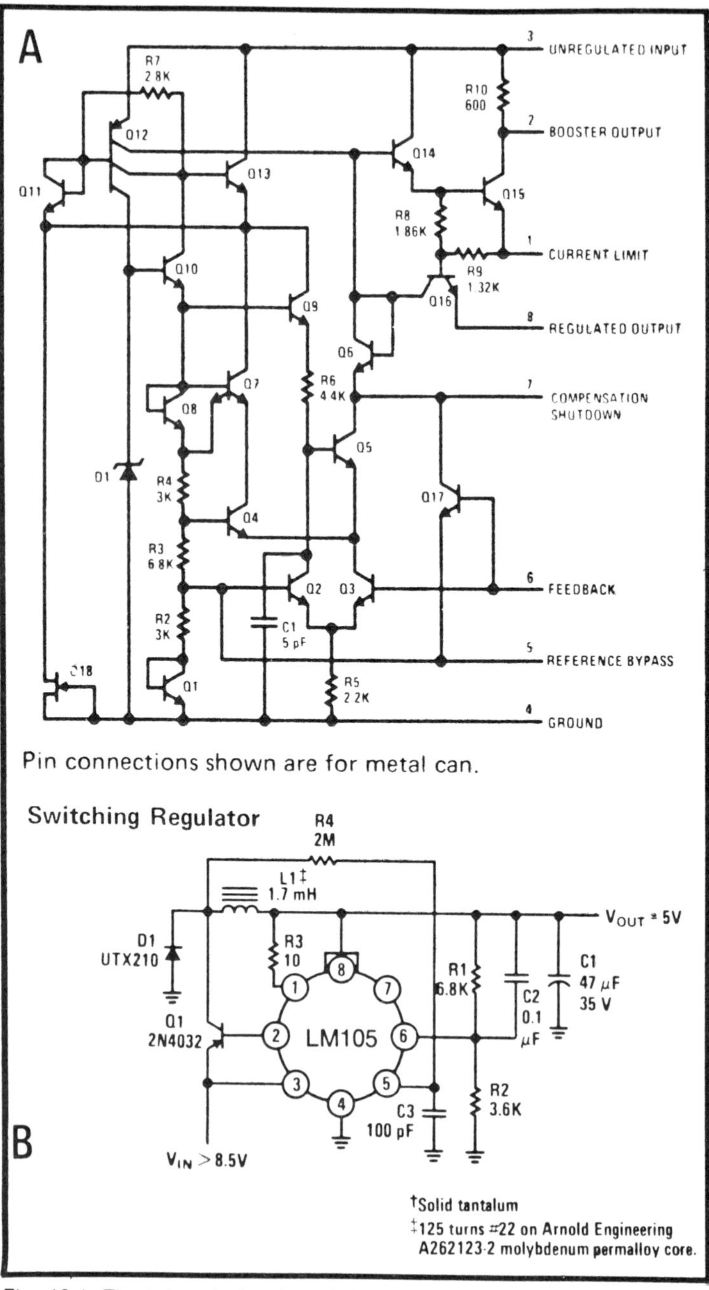

Fig. 13-1. The internal circuitry of the LM105 voltage regulator and its adaptation to switching-mode operation. (Courtesy National Semiconductor Corp.)

thermal stability, packaging, and economic considerations. Zener diode D_1 develops the reference voltage, and the current for D_1 is derived from one of the collectors of Q_{12}. The use of a constant-current source for the operation of D_1 is the first step in producing a very stable voltage reference. The zener voltage across D_1 is not directly utilized; rather, it is buffered by Q_{10}. The output of Q_{10} is applied to the series-connected array comprising Q_1, R_2, R_3, R_4, and Q_8. (Diode-connected transistor structures such as Q_1 and Q_8 appear commonly in monolithic circuits.) In this instance, the overall effect of the series circuit is to provide a reference voltage of 2.2V to the base of Q_2—with a nearly zero coefficient of temperature. It can be seen that Q_2 is part of the comparator, consisting of Q_2 and Q_3. Transistor Q_5 boosts the voltage from the output of the comparator, and Q_{14} provides drive for series-pass transistor Q_{15}. Most of the voltage amplification is developed in Q_5 because its collector load is essentially the high-impedance, constant-current source from the collector of transistor Q_{12}. Other active devices are involved in auxiliary functions; for example, the field-effect structure (Q_{18}) provides initial base bias to Q_{12} during startup, and thereafter Q_{12} receives its base bias from Q_9. Transistor Q_{16} participates when current limiting is implemented in the external circuit; it chokes off base current to Q_{14}, thereby allowing pass transistor Q_{15} to deliver no more than a predetermined current to the load. Transistor Q_{17} prevents paralysis of the voltage comparator, a situation arising from saturation of Q_3 when overdriven.

All of the essentials of a switching-type regulator are present in the LM105. Specifically, we have an excellent voltage-reference source, a voltage comparator, error-voltage amplification, and a series-connected output transistor. This leads naturally to the switching regulator shown in Fig. 13-1B. The external transistor (Q_1) extends the current output to several hundred milliamperes. This circuit constitutes a basic approach for many switching regulators. It should be appreciated that the current-boost technique provided by the external transistor may be extended by cascading progressively larger transistors. Of course, the current-carrying capacity of L_1 and D_1 must also be appropriately increased. Hundreds of amperes may be precisely and efficiently regulated by this cascading technique.

There are two feedback paths in the circuit of Fig. 13-1B, as well as in most self-oscillating regulators. One of these, a *negative* feedback path, brings about regulation of the output voltage by minimizing the error signal from the comparator. Although the action is somewhat different, the effect is very similar to that of a series-pass linear supply, and the actual circuit connections are also similar in the two types of regulators. The negative feedback is obtained through the output sampling network, R_1, R_2, and C_2, and is applied to pin 6, the inverting input of the comparator. Capacitor C_2 provides a low-attenuation feedback path for ripple and noise, thereby reducing the level of these AC components on the output line. (The ripple primarily affected is that contributed by the unregulated supply rather than that corresponding to the switching rate.)

The second feedback is *positive* in nature and causes the self-oscillation of the system. The positive feedback stems from the connection of R_4 from the input of the inductor to the noninverting input of the comparator, pin 5. Resistor R_4 also establishes the peak-to-peak ripple voltage because its value governs the hysteresis of the comparator. Regulation is achieved by varying the switching duty cycle, or by changing the switching rate or the duration of *on* time. The voltage level of the regulated DC output can be adjusted by means of the R_1/R_2 ratio, as in a linear regulator.

The basic concepts embodied in this switching-type regulator are similar to many others. Various modules do not all coincide in pin numbers; for example, in the LM723 type, the voltage-reference source is not internally connected to the voltage comparator; rather, it appears at a pin, where it may be used for various circuit purposes.

NEGATIVE-VOLTAGE SWITCHING REGULATORS

Although a voltage regulator such as the LM105 is intended to be used for a positive output polarity, it can also be adapted to deliver a negative output. Still better results and easier implementation may be obtained by using ICs specifically designed for negative output voltages. An example of such a voltage regulator is the LM104, shown in Fig. 13-2. Two negative-output switching regulators are shown in Fig. 13-3. The circuit philosophy and applications are very similar to the LM105.

In the high-current circuit of Fig. 13-3B, the extended performance is derived from several modifications. Both the inductor and the free-wheeling diode must have greater current capability. Note that the reference supply terminal is no longer returned to the unregulated input, but is connected to the base of output transistor Q_2. This stratagem is used in order to prevent pin 5 from becoming more than 2V positive with respect to pin 3—a condition which would adversely affect the operation of the LM104. Although these modifications accomplish its primary objective, they may produce an undesirable side effect from some applications, since the line regulation is degraded by the unregulated input voltage being injected into the voltage reference source. Fortunately, this effect can be eliminated by inserting a 0.01 μF capacitor in series with positive-feedback resistor R_6. The capacitor is large enough to have negligible effect upon the comparator hysteresis at the oscillation frequency, but blocks the DC component of the feedback.

As current capability is increased, the need for fast-responding semiconductor devices becomes more pressing. Not only is more heat evolved from slow elements, but at high current levels, the severity of circuit disturbances from high peak currents and chassis loop currents mounts rapidly. Low-frequency workhorse transistors and free-wheeling diodes should not be used. The junction charge storage of ordinary rectifying diodes not only results in high dissipation in the diode itself, but constitutes a short circuit for the switching transistor when it first turns on. (This can hardly be classified as "free-wheeling.") As emphasized in previous chapters, the switching-type supply requires the *coordinated* action of proper devices and components. A single bargain-basement item can harmfully degrade the chain of events required for high efficiency and clean operation.

1 ADJUSTMENT
2 REFERENCE
3 REFERENCE SUPPLY
4 COMPENSATION
5 UNREGULATED INPUT
6 CURRENT LIMIT
7 BOOSTER OUTPUT
8 REGULATED OUTPUT
9 GROUND
R16 15K
R15 15K
R14 2K
R13 1K
R17 7.5K
R4 14K
R1 2.3K
R2 7.8K
R3 7K
R5 2.7K
R9 80K
R10 3K
R11 10K
R8 2.2K
D1 6.3V
C1 5 pF
C2 15 pF
Q1 Q2 Q3 Q4 Q5 Q6 Q7 Q8 Q9 Q10 Q11 Q12 Q13 Q14 Q15 Q16 Q17 Q18 Q19 Q20 Q21 Q22 Q23 Q24
Fig. 13-2. Internal circuitry of the National Semiconductor LM104.

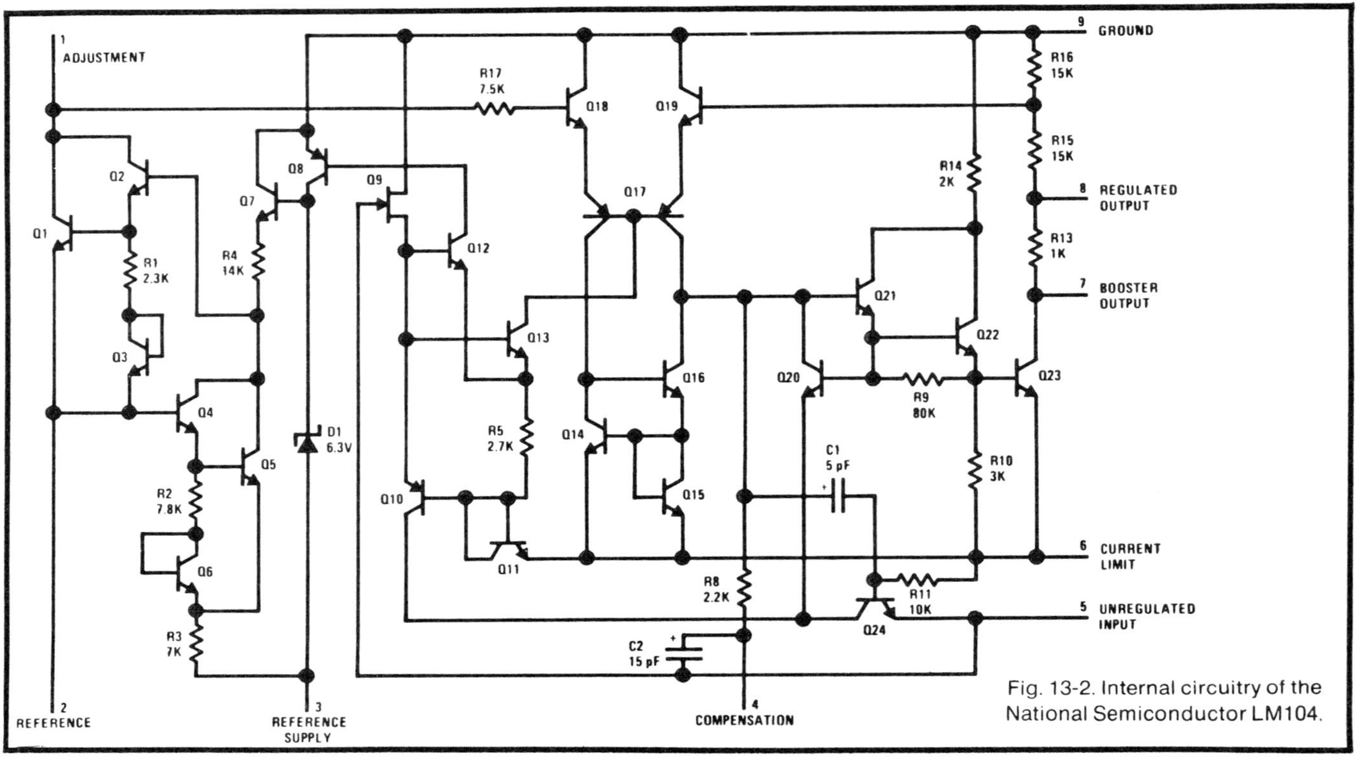

212

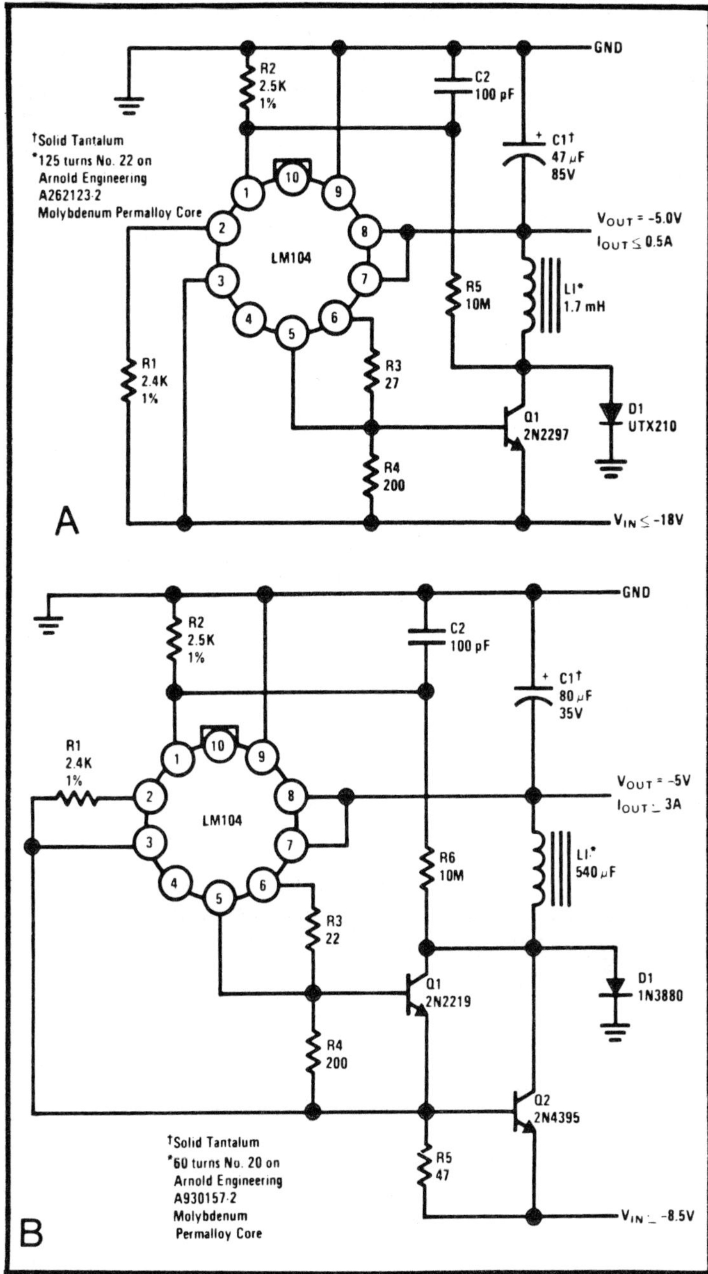

Fig. 13-3. Negative-voltage switching supplies. (A) Single boost transistor design provides −5V output. (B) Two boost transistors provide increased output current. (Courtesy National Semiconductor Corp.)

213

PROTECTION TECHNIQUES

Many a designer has breathed a premature sigh of relief after debugging and refining a newly developed switching regulator. Generally, there is an awareness that some sort of protection against short circuits and overloads must be incorporated, but this is often handled as an afterthought—a minor bit of last-minute patchwork. But a protective function is anything but a trivial problem. A common first approach is to limit the current through the switching transistor by starving its base of forward drive. Since the switching transistor tends to stop switching when the output of the regulator is shorted, the dissipation will then be excessively high. This type of current limiting is often applied to linear regulators, but it will not provide the requisite protection here. Either the switching operation must continue or the switching transistor must be turned off.

Many hours of experimentation can probably be eliminated by utilizing one of the two schemes shown in conjunction with LM104 regulators in Fig. 13-4. In method (A), transistor Q_3 senses the load-current-induced voltage developed across the small resistance R_9. When Q_3 is driven into conduction because of excessive load current, it exerts control over the internal error amplifier via the connection made to terminal 8. The schematic diagram of the LM104 should be studied to see how this is accomplished, but the significant point is that the control of the regulator by switching transistor Q_2 is relinquished to Q_3 in the event of an overload or short at the output of the supply. In essence, transistor Q_3 becomes a series-pass element to maintain regulated voltage at terminal 8, thereby keeping the overall switching regulator in its oscillating mode. At the same time, transistor Q_3 does not suffer the consequences of a short-circuited output, because of the isolation provided by R_{13}. When the short or overload is removed, Q_3 turns off again and the system reverts to normal operation as a switching-type voltage regulator. Thus, no manual reset is required.

In the method shown in Fig. 13-4B, excessive load current turns off the switching process, with switching transistor Q_2 being held in its *off* state of conduction. There is negligible

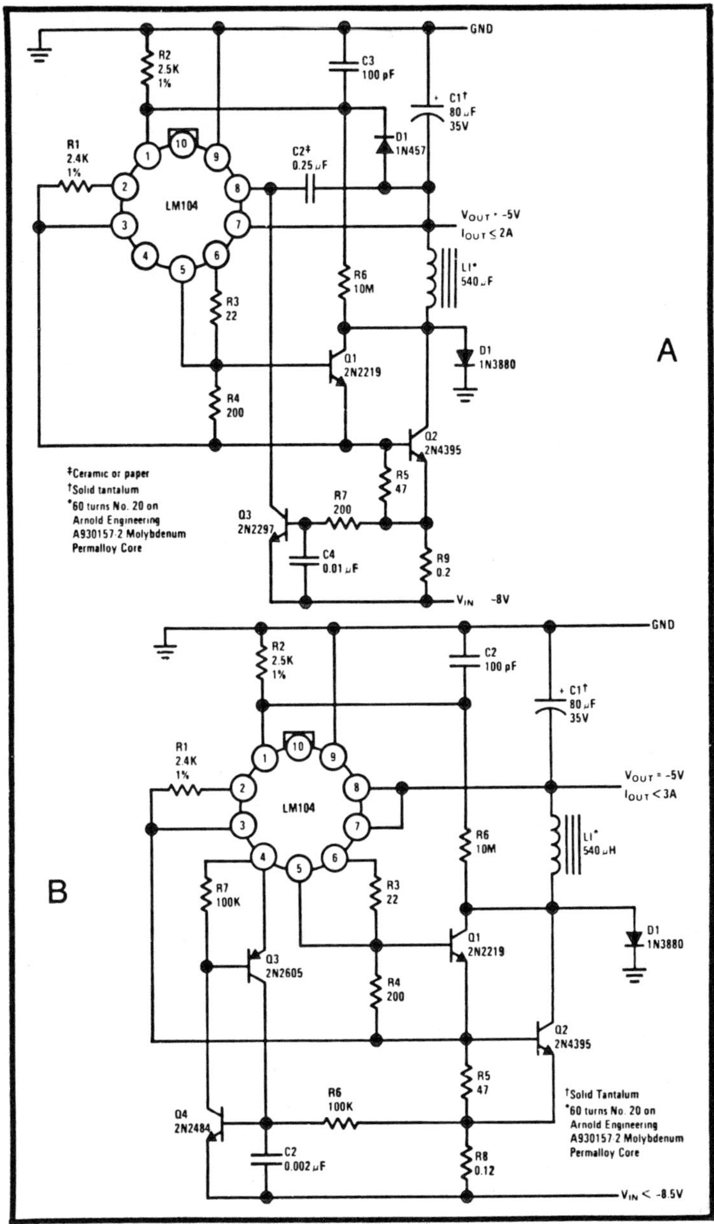

Fig. 13-4. Two overload protective techniques for the LM104 switching circuit. (A) Protection is imparted by current limiting; oscillation is maintained. (B) Protection results from a shutdown of the switching process with the switching transistor held in its off state. (Courtesy National Semiconductor Corp.)

dissipation in this transistor, even with a short-circuited load. Although this scheme provides the ultimate in safety, operation must be restored by *manually* interrupting the unregulated input supply. The protective action operates as follows:

Transistors Q_3 and Q_4 form the equivalent of an SCR, with the base of Q_4 constituting the gate. Excessive load current will develop sufficient voltage across sensing resistance R_8 to fire the simulated SCR. Then Q_3 and Q_4 will latch each other in saturated conduction, and this action through terminal 4 turns off the regulators, by depriving transistor Q_{21} of forward bias. Here is an example of versatility, since terminal 4 not only is ordinarily used for frequency compensation, but serves well for a shutdown technique. These protection circuits can be readily adapted to other regulator ICs.

A SWITCHING/LINEAR COMBINATION

If we were not concerned about efficiency and such related factors as heat removal, cost, and packaging dimensions, there would be no dissatisfaction at all with the linear regulator. The proponents of regulation by dissipation have investigated many avenues whereby excess dissipation could be reduced. For example, there have been designs for automatic switchover from series to shunt regulation at strategic load-current values. A more popular approach is to employ a second feedback loop, which functions to maintain a near constant drop across the series-pass transistor. The prime source of inefficiency in these rheostat-like regulators resides in the voltage differential between unregulated input and regulated output. The dissipation in the pass transistor will be minimal if the unregulated input voltage can be maintained just high enough to insure proper regulator action. Further, there can be no objection to regulating the DC input voltage, since such preregulation must add to the overall performance of the regulating system.

This naturally leads to the use of a switching regulator followed by a linear regulator. Such a system is illustrated in Fig. 13-5. Two LM105 modules are used in a configuration which, despite its dual-mode operation, retains the basic

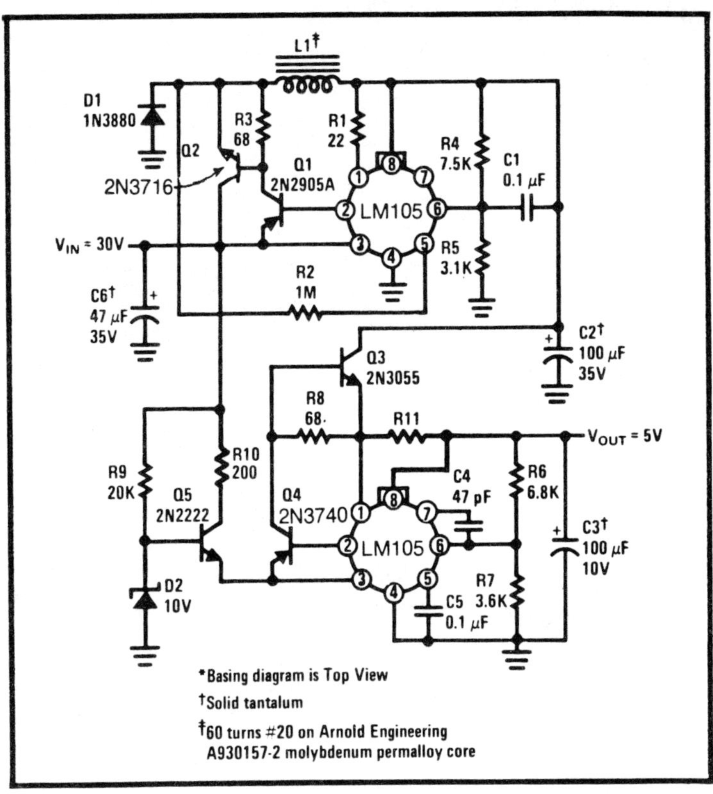

Fig. 13-5. Combined switching and linear supply. (Courtesy National Semiconductor Corp.)

circuit simplicity that these modules generally provide. Despite the 25V differential between the nominal input voltage and the regulated 5V output, the efficiency of this regulating scheme can range between 60 and 70%. The network formed by Q_5, D_2, R_9, and R_{10} comprises a constant-voltage source, which isolates the linear regulating circuitry from noise and transients on the unregulated DC line. Many commercial power supplies use such a scheme to achieve high operating efficiency, together with low output ripple and fast transient response. To scale a design for greater output current the current capabilities of both the switching and the linear regulators must be increased. (This is obvious enough, but still a point to be watched when ordering the highest current rating from any family or group of commercial power supplies.)

SIMPLE SWITCHING REGULATOR

A natural reaction to a diagram devoid of the usual complexities is that marginal performance must ensue. Yet, the straightforward switching regulator shown in Fig. 13-6 does not involve tradeoffs with basic power supply parameters. Inspection of the schematic diagram of the LM341 shows that despite the fact that only three connections emerge from this IC, the internal circuitry is like that of the high-performance LM104. Incorporated in the design of the LM341 is complete protection, including both current limiting and thermal shutdown.

The LM341 was initially intended for use as a point-of-load or on-card linear regulator. The use of several regultors physically situated in close proximity to the loads, deserves serious consideration in large installations. Too often, the splendid DC and AC regulating characteristics of a sophisticated and costly source of operating power are considerably degraded by the resistance and inductance of the connecting leads. It may prove advantageous to relax the specifications on the main power supply and to use small point-of-load regulators where necessary. Such localized regulation has usually been accomplished by linear regulators; however, there are instances where a small, easily applied switching regulator would be very desirable. Although we tend to ignore low efficiency at low power levels, the cumulative dissipation of *many* small linear regulators could prove undesirable. The three-terminal switching circuit may provide an acceptable solution, but much depends upon the nature of the load, primarily its susceptibility to high-frequency ripple and RFI resulting from the switching process.

A point of caution is in order: the tab of the package is electrically identical with terminal 3. In the switching circuit shown in Fig. 13-6, terminal 3 is *not* directly grounded. Therefore, it would not be permissible to use a mounting method in which the tab makes electrical contact with the chassis in most situations. Similar reasoning pertains to the PNP switching transistor—its collector and case (if metallic) must not be inadvertently grounded.

schematic and connection diagrams

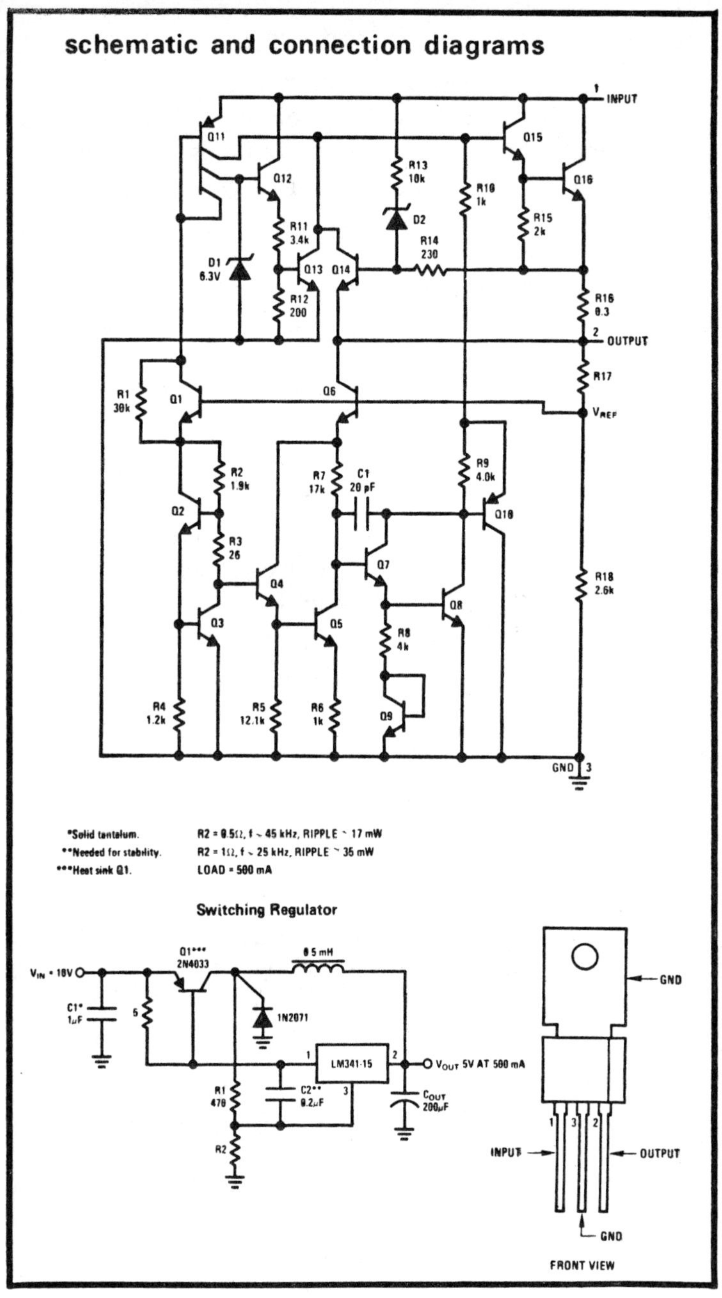

Fig. 13-6. Switching regulator utilizing the LM341 three-terminal IC. (Courtesy National Semiconductor Corp.)

PROTECTED SWITCHING REGULATOR

The switching regulator shown in Fig. 13-7 involves no significant differences from those already discussed. The basic principles of operation are the same as those already described for self-oscillating switching regulators. However, the use of a new semiconductor device as the switching element gives this regulator a new dimension of performance. The LM195, shown in simplified circuitry in Fig. 13-7, has only three terminals and physically resembles a conventional power transistor. And it is used in circuits very much like ordinary transistors. But there the resemblance ends, for this product is in a class by itself—it cannot be considered as just another power transistor.

This monolithic power module includes provisions which limit its current and power output and shut it off in the event of excessive temperature rise. Once properly incorporated in a circuit, the LM195 is virtually immune to destruction from any type of overload. Such behavior is particularly welcome for application to switching-type power supplies, where reliable overload protection is considerably more difficult to achieve.

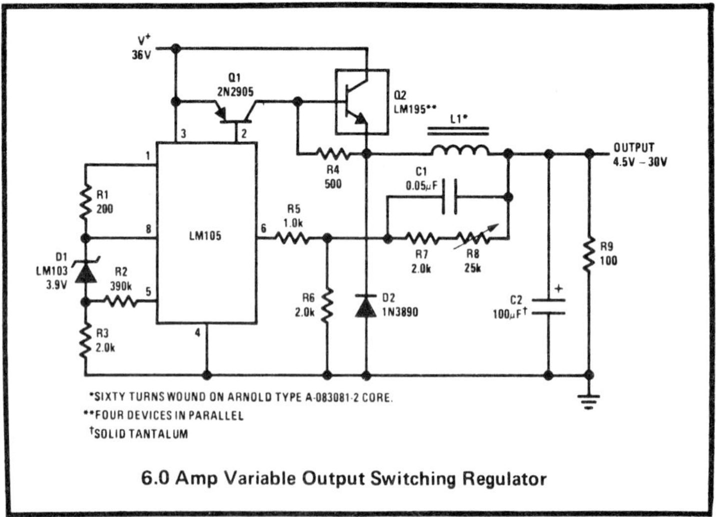

Fig. 13-7. Self-protected switching regulator utilizing the LM195 monolithic power transistor. The difficult problem of overload protection is neatly solved through the use of this simulated power transistor. (Courtesy National Semiconductor Corp.)

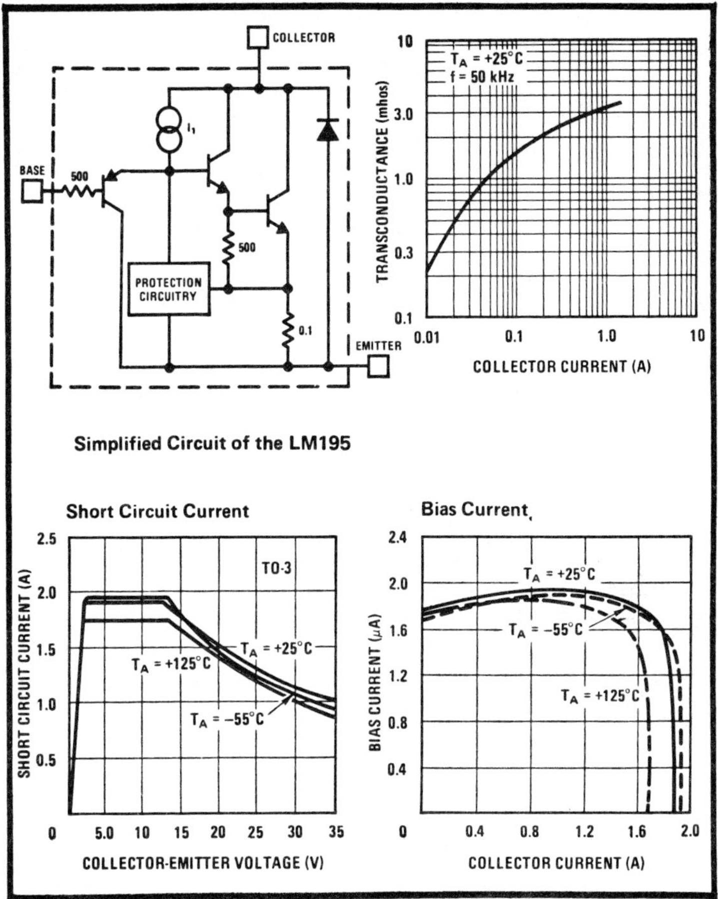

Simplified Circuit of the LM195

Fig. 13-8. The LM195 monolithic power transistor. The simplified circuit shows the three "transistor" terminals. Curves display high transconductance range at 50 kHz and its self-protection characteristics. Power dissipation is limited to 40W.

In conventional approaches, it is not uncommon to encounter difficult problems of interaction between the switching circuit and the added protective circuitry. This is an area where compromise is hardly desirable.

In addition to its unique self-protection features, the LM195 is a high-performance device at the currents and frequencies which are useful for switching regulators. The high transconductances shown in Fig. 13-8 were measured at 50 kHz. This contrasts favorably with many ordinary power

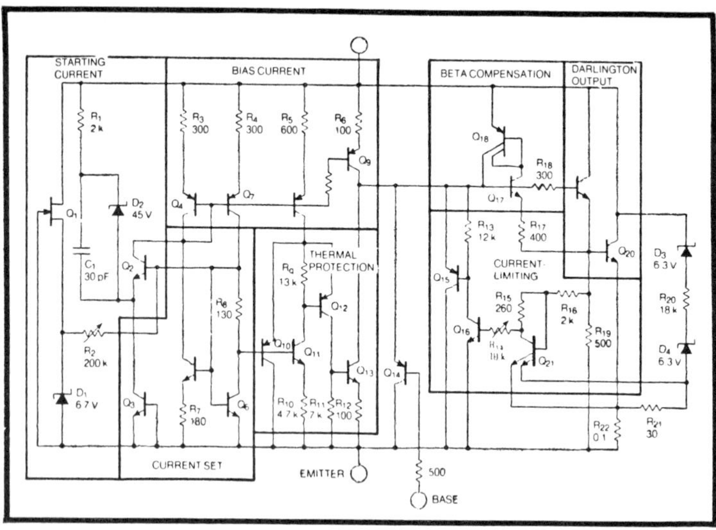

Fig. 13-9. Internal circuit of LM195 monolithic power transistor. In addition to performing the role of a power transistor, this unique device also confers automatic protection against its own destruction from overloads and from excessive temperature rise. (Courtesy National Semiconductor Corp.)

transistors which require high driving power. The inherent protective features of the LM195 are also depicted. Not shown is the fact that the "base" of this device can be driven up to 40V without causing breakdown, and this too facilitates the design of a switcher. The actual monolithic circuitry is presented in Fig. 13-9. This exemplifies the clear superiority of ICs for complex functions, in which discrete devices would be expensive, bulky, and impractical.

A 100W SWITCHING REGULATOR WITH SHUNT CHOPPER

Otherwise known as a *flyback* regulator, this circuit is most commonly seen at low power levels. Over about 150 watts, such a system becomes uneconomical because of the large peak currents the switching or chopping device must withstand. However, good performance is readily achieved from this unique technique, and it can deliver regulated DC at a *higher* voltage level than that obtained from the unregulated supply. For some applications, this is a compelling feature.

The circuit shown in Fig. 13-10 is a straightforward shunt switcher. It is essentially a constant-frequency, variable-pulse-width scheme. In order to accomplish this operational mode, an oscillator and a duty-cycle-controlled multivibrator are used. The oscillator consists of the unijunction transistor circuit, configured about the 2N1671B UJT. This circuit functions as a simple relaxation oscillator at a nominal frequency of 9 kHz. The next three bipolar transistors, from right to left, comprise a monostable multivibrator with voltage-controlled pulse width. The control voltage is sampled from the output of the regulator; as the load increases, the shunt switching transistor remains on longer in order to store more energy in the inductor. When the transistor turns off, the released energy from the inductor manifests itself as higher output voltage. Because of the error-voltage feedback and the comparison made with the reference voltage, the restoration of the output voltage is nearly exact. This constitutes regulation and is similar to the action that takes place in conventional series-type switchers.

The 2N3706 transistor, at the extreme left, functions as a driver for the shunt switch. The shunt switch is a DTS-1020

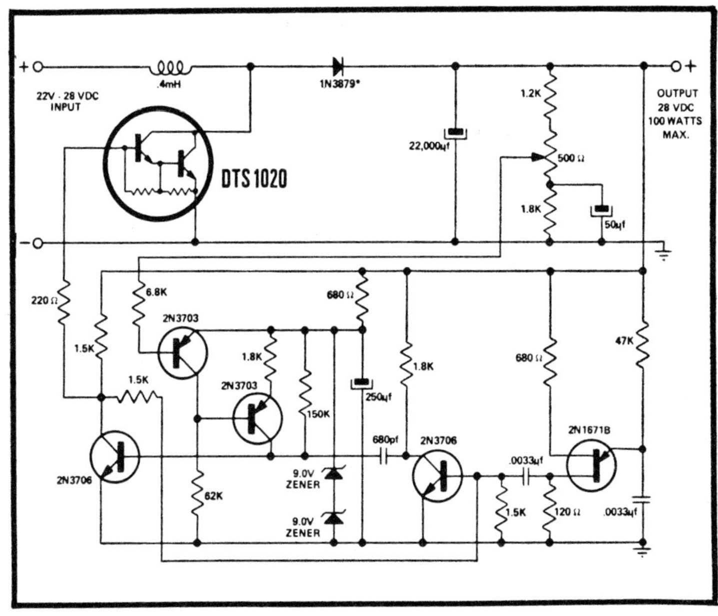

Fig. 13-10. 100-watt flyback switching regulator. The 0.4 mH choke consists of 124 turns 17 AWG wire on Arnold B079024-3 core. (Courtesy Delco Electronics Division)

Darlington silicon power transistor. The inductor uses a low mu, powdered-iron core in order to attain constant inductance throughout its nominal current range, as well as a "soft" saturation characteristic at high currents.

The efficiency quickly rises to the vicinity of 85% at one-third rated load, and it peaks out somewhat higher than this at around 80W output. This is quite respectable performance, considering that this type of regulator has not received the developmental effort expended on the series-switching type. Both regulation and ripple are less than 1% at the rated 100W output level.

HIGH-VOLTAGE SWITCHING REGULATOR

The relatively simple circuit of Fig. 13-11 exemplifies the performance that is now attainable with high-technology transistors and diodes. Both the voltages and the power levels indicated in the chart were in the realm of fantasy a few years ago—or were priced beyond reach of all but military and space-oriented projects. The use of simple discrete transistors in the control section caters to a school of thought which sees no need of greater sophistication at high power levels. This, of course, is a designer's prerogative, but it cannot be disputed that this control technique is rugged. It is perhaps readily made immune to the effects of large circulating currents, often induced by external equipment. In any event, full-load efficiencies as high as 92% are obtainable. This, together with 0.6% for combined line and load regulation and 0.75V peak ripple at full load, should dispel any doubts about performance.

In Fig. 13-11, the two 2N2711 transistors, Q_2 and Q_3, are connected as a differential amplifier which, together with the 1N751 zener diode, functions as a straightforward voltage comparator. Operation is constrained to its linear region. Inasmuch as we are dealing with a switching-type regulator, rather than a linear type, the quantity sampled by the comparator can best be thought of as voltage ripple, rather than as a sustained DC level. The comparator develops an error signal which ranges above and below the reference voltage. Because of the proportionate amplification by the

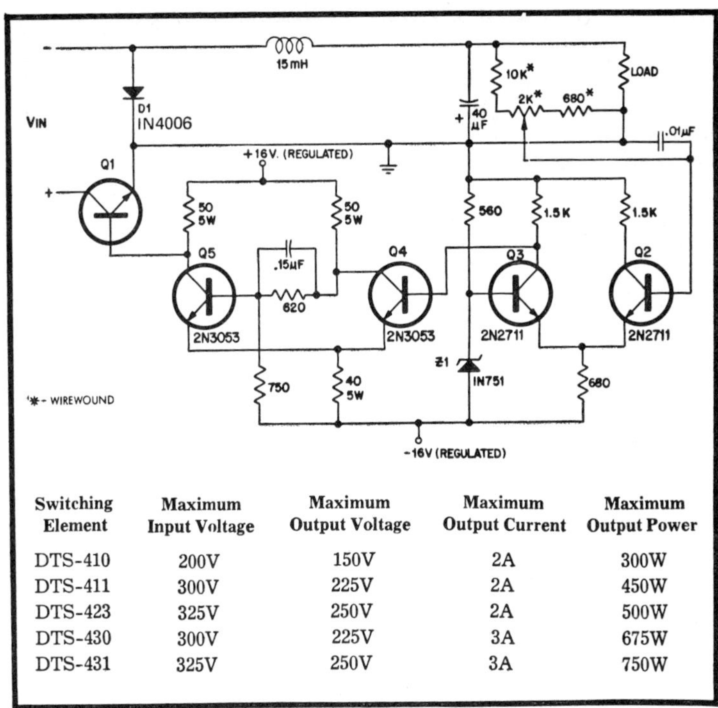

Switching Element	Maximum Input Voltage	Maximum Output Voltage	Maximum Output Current	Maximum Output Power
DTS-410	200V	150V	2A	300W
DTS-411	300V	225V	2A	450W
DTS-423	325V	250V	2A	500W
DTS-430	300V	225V	3A	675W
DTS-431	325V	250V	3A	750W

Fig. 13-11. A versatile high-voltage switching regulator. (Courtesy Delco Electronics Division)

comparator, this error signal retains its triangular waveshape.

The next stage, comprising transistors Q_4 and Q_5, is a Schmitt trigger. This stage produces variable-duration pulses in response to the amplitude variations delivered by the comparator; the overall control circuitry thus behaves as a pulse-width modulator. Switching transistor Q_1 is directly driven by the Schmitt trigger stage. The final act in this sequence of events is output-voltage regulation by control of the switching duty cycle. The nominal frequency of self-oscillation of this supply is in the vicinity of 7 kHz.

ALL-DISCRETE 1000W SWITCHING REGULATOR

The circuit shown in Fig. 13-12 provides a regulated output of 500V at 2A. The configuration is unencumbered; its simplicity is suggestive of circuit techniques commonly used at lower power and voltage levels. The fixed switching

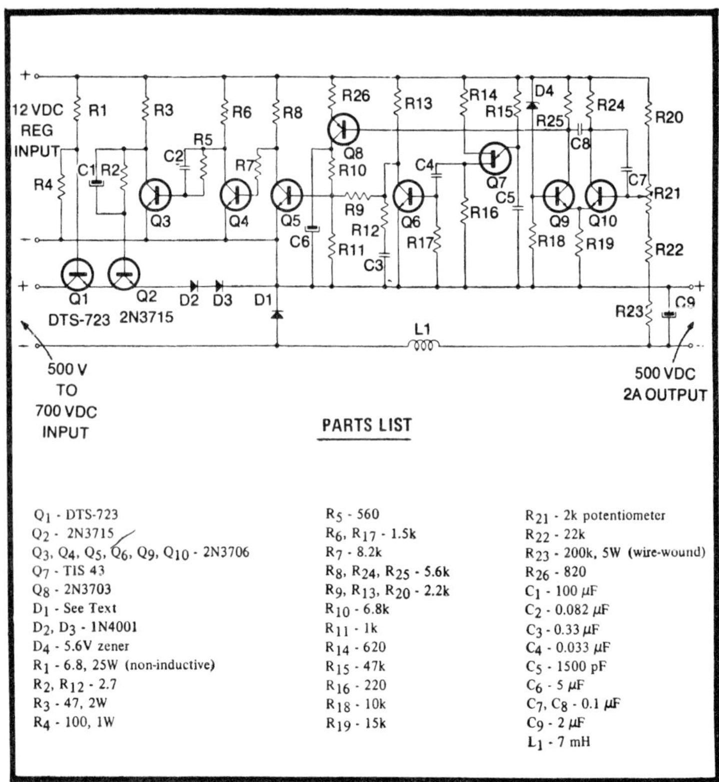

Q_1 - DTS-723
Q_2 - 2N3715
Q_3, Q_4, Q_5, Q_6, Q_9, Q_{10} - 2N3706
Q_7 - TIS 43
Q_8 - 2N3703
D_1 - See Text
D_2, D_3 - 1N4001
D_4 - 5.6V zener
R_1 - 6.8, 25W (non-inductive)
R_2, R_{12} - 2.7
R_3 - 47, 2W
R_4 - 100, 1W

R_5 - 560
R_6, R_{17} - 1.5k
R_7 - 8.2k
R_8, R_{24}, R_{25} - 5.6k
R_9, R_{13}, R_{20} - 2.2k
R_{10} - 6.8k
R_{11} - 1k
R_{14} - 620
R_{15} - 47k
R_{16} - 220
R_{18} - 10k
R_{19} - 15k

R_{21} - 2k potentiometer
R_{22} - 22k
R_{23} - 200k, 5W (wire-wound)
R_{26} - 820
C_1 - 100 μF
C_2 - 0.082 μF
C_3 - 0.33 μF
C_4 - 0.033 μF
C_5 - 1500 pF
C_6 - 5 μF
C_7, C_8 - 0.1 μF
C_9 - 2 μF
L_1 - 7 mH

Fig. 13-12. One-kilowatt switching regulator featuring simple design. (Courtesy Delco Electronics Division)

frequency is 13 kHz, and is established by unijunction transistor Q_7. Transistor Q_6 functions as a sawtooth generator, being triggered from the unijunction stage. The sawtooth waveform is modulated by DC voltage from Q_8, which in turn is controlled from the output of the error-sensing differential amplifier, comprised of transistors Q_9 and Q_{10}. The consequence of the DC modulation is that the DC level of the sawtooth wave appearing at the base of transistor Q_5 varies with the error signal. The error signal is generated by the input differential amplifier, Q_9 and Q_{10}, as the difference between the sampled output voltage and the reference voltage established by zener diode D_4. Transistors Q_5 and Q_4 function as a pulse-width modulator, because the varying DC level of the sawtooth wave causes Q_5 to respond to varying portions of the sawtooth. Because of saturation in the pulse-width modulator, the output consists of a rectangular waveform with *on* and *off* times dependent upon the DC component of the sawtooth driving wave. Except for power level, the pulse-width-modulated wave from Q_4 is now satisfactory for controlling the duty cycle of the switching process.

Transistor Q_3 is a predriver. Transistor Q_2 is the driver stage for switcing transistor Q_1. The duty cycle of Q_1 is varied in such a manner as to produce a near constant output voltage. The combination of Q_2 and Q_1 is of special interest—Q_1 is connected in the common-base configuration with its emitter driven from Q_2, which is a fall-time speedup technique. When Q_1 is turned off, its emitter—base section is avalanched, quickly depleting the stored base charge.

The minimum requirements for free-wheeling diode D_1 are 800 PIV, 3A peak repetitive forward current, and 0.5 μsec minimum reverse-recovery time. The combination of a fast diode and rapid turnoff switching transistor makes possible an operational efficiency in the 90% range. The load regulation is on the order of $\pm 0.5\%$, with most of the change occurring at loads exceeding 1.5A. The circuit could be scaled for operation at 20 kHz, but the output should then be restricted to 500W.

5 kW SWITCHING REGULATOR

Any circuit-oriented worker in electronics is bound to have his interest stimulated by the unusual topography of the 5000W switching regulator shown in Fig. 13-13. Not only does the schematic depart from the beaten path, but the operating parameters are also uncommon. For example, despite the tremendous power capability of this switcher, the efficiency is in the vicinity of 95% from 3 kW to full output, and for all loads greater than 5000W, the efficiency exceeds 90%. Line regulation is better than 0.1% over the full input-voltage swing from 500 to 700V. And load regulation is less than 0.5% when the resistive load is changed from 500W to 5 kW. This combination of outstanding performance parameters reflects the excellence of both the Delco switching transistors and the National Semiconductor quad amplifier used in the control section.

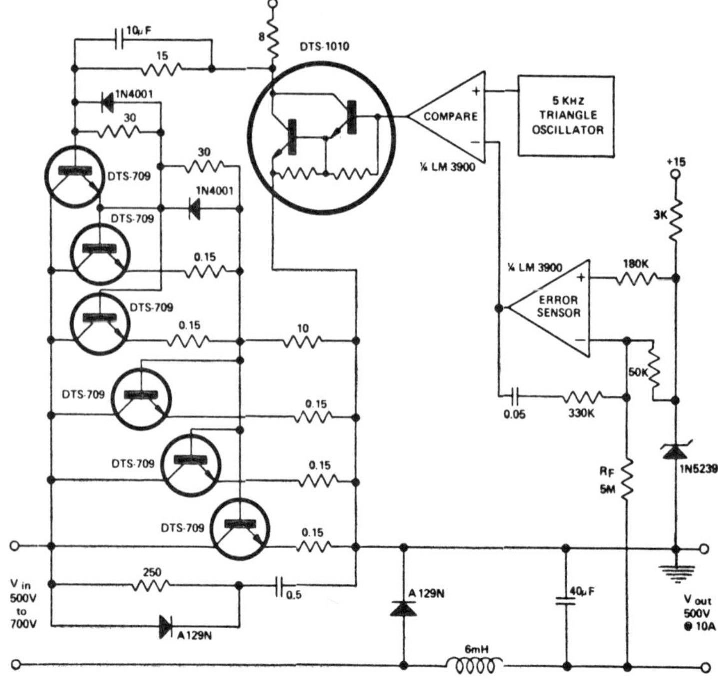

Fig. 13-13. Five-kilowatt switching regulator. (Courtesy Delco Electronics Division)

The LM3900 quad amplifier differs from conventional operational amplifiers; a more detailed look into its circuitry and its applications relevant to the regulator is provided in Fig. 13-14. The *current-mirror* principle enables simulation of a differential-amplifier input circuit; the input-biasing current, however, is inordinately low. Operation is generally restricted to AC signals, but all of the benefits of direct coupling are retained. This configuration develops high open-loop gain, with a relatively low number of active elements. Implementation is made convenient because of internal frequency compensation (for unity gain) and by virtue of the single power supply requirement.

The regulator contains a "snubber" circuit, comprising diode AT129N, the 250Ω resistor, and the 0.5 μF capacitor. This network safeguards the three parallel-connected DTS-709 switching transistors, maintaining their operation within the

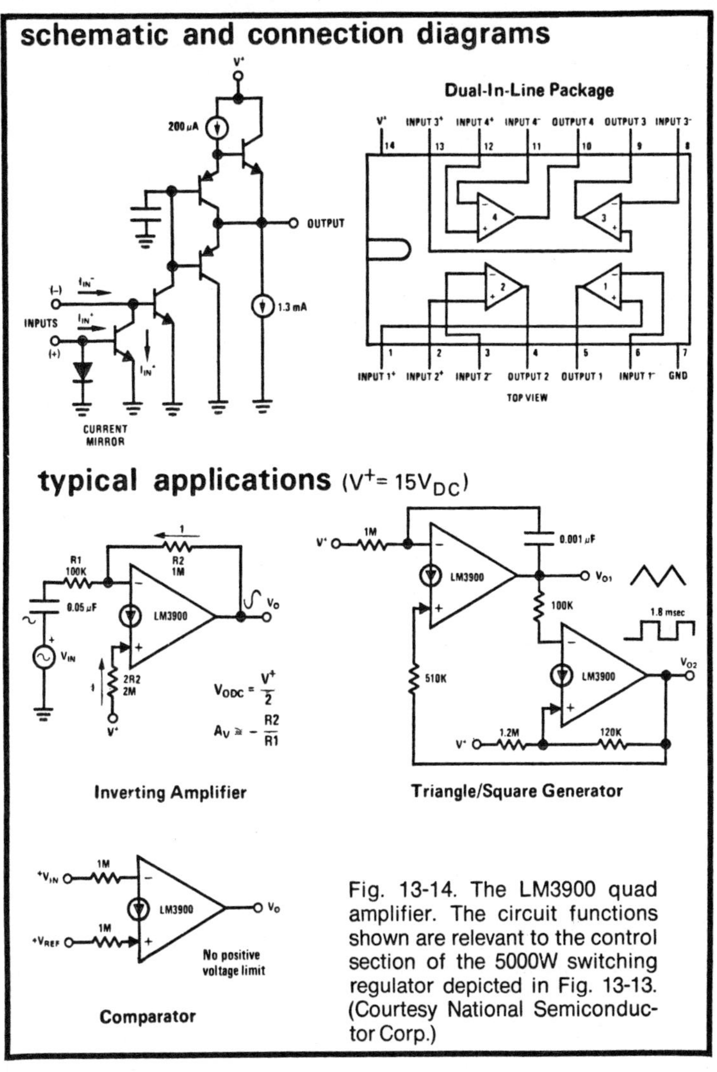

Fig. 13-14. The LM3900 quad amplifier. The circuit functions shown are relevant to the control section of the 5000W switching regulator depicted in Fig. 13-13. (Courtesy National Semiconductor Corp.)

bounds of reverse bias, second breakdown. These three output transistors are driven by two parallel-connected DTS-709s, and they, in turn, receive drive from a single DTS-709. The array of DTS-709 transistors comprises a triple-Darlington arrangement. The predriver, an integral Darlington DTS-1010, provides voltage amplification of the width-modulated pulses received from the comparator. Thus, this regulator is actually quite straightforward. Its dollar cost per kilowatt would certainly meet low-cost requirements. Even greater economy can be obtained by operating from a 480V, 3-phase, full-wave-rectified line which uses minimal filtering.

ADVANCED-DESIGN SWITCHING REULATOR

The switching regulator shown in Fig. 13-15 merits discussion because it embodies a combination of the features which have appeared singly in a number of other circuits. One of the features is not only unique, but represents a significant advance in the art of switching-type power supplies. This *load-line shaping* network comprises the elements enclosed in the shaded area. Its function will be described, but first,

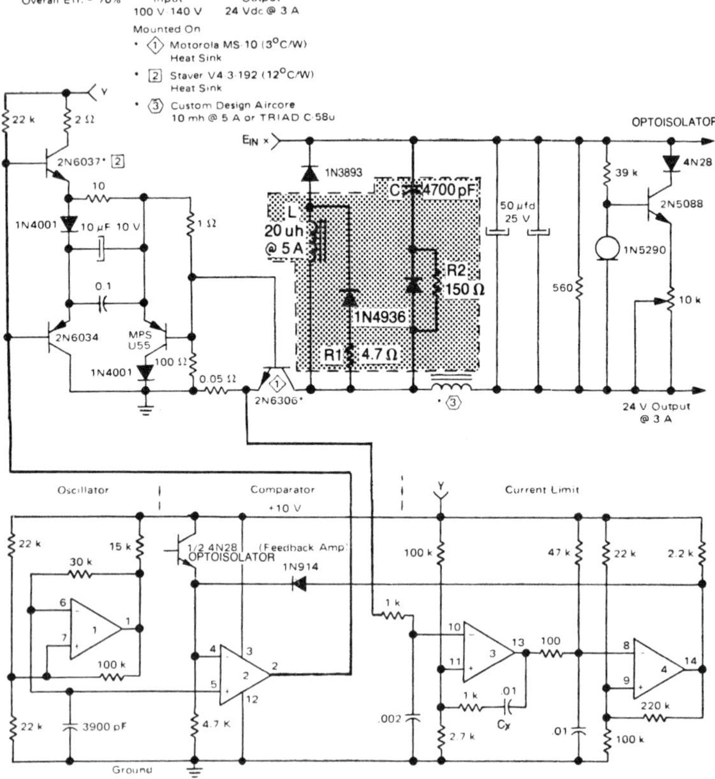

Fig. 13-15. Switching regulator with advanced design features. Regulator incorporates load line shaping (shaded area), optoisolator feedback, quad-function IC package, overcurrent protection, and direct operation from the power line. The unregulated and auxiliary power supplies are shown in Fig. 13-17. (Courtesy Motorola Semiconductor Products, Inc.)

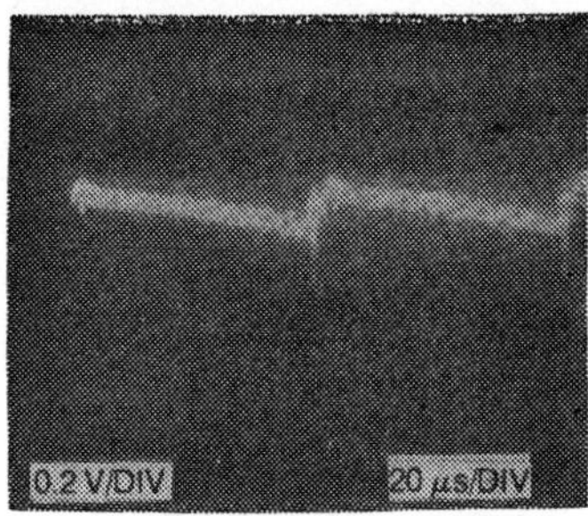

Fig. 13-16. Output noise cleanup resulting from load-line shaping network. Not only is the switching transistor protected from excursions into dangerous operating regions, but the generation of switching spike is drastically reduced. (A) Regulator of Fig. 13-15 without load-line shaping network. Note the normal switching spikes. (B) Same measurement, but with load-shaping network connected in circuit. (Courtesy Motorola Semiconductor Products, Inc.)

however, it would be appropriate to list the features of this regulator:

- Load-line shaping—this is an important evolutionary step in switcher teconology. Suffice it to say, for the moment, that much cleaner switching is accomplished with the addition of this network. Figure 13-16 dramatically illustrates the noise reduction attained through the use of this technique.
- Direct operation from the power line—there is no bulky 60 Hz transformer; and despite the 20 kHz operating frequency, there is no inverter.
- An optoisolator is used in the feedback loop.
- Short-circuit protection and current limiting are provided.
- All control logic, including the overcurrent protection, is provided by a quad-function IC package, so that the circuitry is relatively unencumbered.

The unregulated power supply, together with the 10V supply for the IC sections, have been separated from the regulator and are shown in Fig. 13-17. Transformer T_1 is very small inasmuch as it has an approximately a 15 voltampere rating. The main power is drawn directly from the power line and rectified by the MDA-980-4 bridge. The connections of these supplies appear on the schematic diagram of Fig. 13-15.

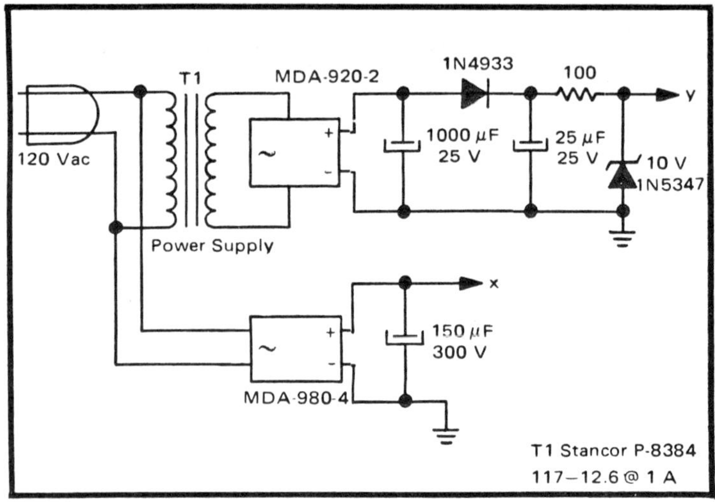

Fig. 13-17. Unregulated and auxiliary power supplies for advanced-design regulator. The MDA-980-4 is a bridge rectifier. Connections are made to the appropriate points labeled X and Y in the schematic of Fig. 13-15. (Courtesy Motorola Semiconductor Products, Inc.)

Since the nominal DC output of the main supply is 160V, and the regulator provides 24V, it follows that the switching process occurs at a low duty cycle, with the *on* time being relatively short. Such a great differential between input and output voltage would be impractical with a linear regulator intended to deliver 24V to a 3A load.

The Basic Principles of Load-Line Shaping

The objective of the load-line shaping network shown in shaded area of Fig. 13-15 is to divert rise-time and fall-time switching losses from the switching transistor. (These losses are designated as areas 1 and 3 in Fig. 8-8.) The energy that would otherwise be dissipated in the switching transistor is stored in the reactances of the shaping network until the full *on* or *off* conditions are attained. Then, this energy is dissipated in the resistors of the shaping network. Inasmuch as energy is dissipated in this way, this technique does not directly increase the overall efficiency of the switching regulator; It makes little difference whether energy is dissipated in the switching transistor or is diverted to resistors, but the indirect advantages are of considerable

232

worth, the SOA requirement of the switching transistor is greatly relaxed, permitting this transistor to operate under more reliable conditions. In a sense, the load-line shaping network can be viewed as an electronic heatsink—heat that would otherwise have to be removed from the switching transistor is released by the shaping-network resistors.

The function of the small 20 μH inductor is to absorb voltage during the finite turnon period of the switching transistor. In effect, the application of DC input voltage E_{IN} is *delayed* until the switching transistor is fully *on*. The diode/resistor combination clamps the induced voltage during turnoff, but permits the inductor current to build up again well within "free-wheeling" time; that is, when the switching transistor is *off*. During the transition time from *off* to *on*, the switching transistor is spared the need to simultaneously operate at high current and high voltage.

The function of the small 4700 pF capacitor is to provide load current during the finite turnoff period of the switching transistor. This current does not have to be provided by the switching transistor. The diode/ resistor combination used with this capacitor permits the capacitor to charge well with *on* time, and to act as a current source during fall-time, when the switching transistor is turning off. During the transition time from *on* to *off*, the switching transistor is also spared the need to simultaneously operate at high voltage and high current.

The design of the load-line shaping network is not at all difficult; however, empirical techniques will usually be needed for final optimization because one is not likely to have accurate knowledge of the rise and fall times of the switching transistor. Fall time, in particular, is greatly dependent upon temperature, base bias, and other factors. The computation of initial values derives from the qualitative description of the network already given. Referring to the schematic diagram of Fig. 13-15, proceed as follows:

Shunt Inductor

$$L = \frac{V_{IN}\, t_R}{I_O}$$

where L = inductance of the shunt inductor in microhenrys. This inductor must not saturate at full load current, I_O. A good margin of safety should be incorporated to prevent this. When feasible, an air-core type can be used to circumvent this problem.

V_{IN} = DC input voltge from the unregulated supply.

t_R = rise time of the switching transistor in microseconds—best obtained by measurement.

I_O = maximum load current in amperes.

Resistor In The Inductive Arm.

$$R_1 = \frac{L}{t_A}$$

where R_1 = resistance in ohms.

L = inductance of the shunt inductor in microhenrys.

t_A = time interval in microseconds (short compared to "free-wheeling" or *off* time.) As a start, t_A can be taken as one-tenth the minimum *off* time involved in the operation of the regulator.

The product $R_1 \times I_O$ yields the voltage overshoot at the collector of the switching transistor because of shunt inductor L. If this is greater than desired, the value of R_1 must be reduced.

Shunt Capacitor

$$C = \frac{I_O\, t_F}{V_{IN}}$$

where C = capacitance in microfarads.

V_{IN} = DC input voltage from the unregulated supply.

I_O = maximum load current in amperes.

t_F = fall time of the switching transistor in microseconds (best obtained by measurement).

Resistor in the Capacitive Arm

$$R_2 = \frac{t_B}{C}$$

where R_2 = resistance in ohms.

t_B = a time interval in microseconds which is short compared to *on* time. As a start, t_B can be taken as one-tenth the minimum *on* time involved in the operation of the regulator.

C = capacitance from previous step.

The quotient of V_{IN}/R_2 yields the current overshoot required from the collector of the switching transistor because of the capacitive arm involving C. If this value is greater than desired, the value of R_2 must be increased.

The Feedback Circuit

The feedback circuit makes use of a triangular wave and an output-derived DC voltage to produce duty-cycle control of the switching transistor. In principle, this is similar to the regulation method used in some of the regulators already discussed. In this regulator, the sampling of the output voltage takes place via the 4N28 optoisolator. Referring to Fig. 13-15, the sampled voltage is initially sensed at the junction of the 1N5290 constant-current diode and the 39K resistor. Inasmuch as the constant-current diode dynamically simulates an extremely high resistance, the dividing effect of conventional sampling networks is avoided; that is, if the constant-current diode were replaced by a resistance, attenuation of the error signal would occur and the feedback loop would be less effective in providing tight regulation.

A voltage level proportional to the sampled output voltage appears at terminal 4 of the comparator, comprised of section 2 of the MC3320 quad-function IC module. There is complete electrical isolation between the regulator output and the comparator, for the optoisolator behaves as a transformer with both DC and AC capability. The other input of the comparator, terminal 5 of section 2, is impressed with a fixed-frequency, constant-amplitude triangular wave. This wave is obtained from the 20 kHz triangle-wave generator configured around section 1 of the quad-function module.

The output of the comparator (terminal 2 of section 2) is a rectangular wave having a duty cycle dependent upon the DC level of the error signal appearing at input terminal 4. The

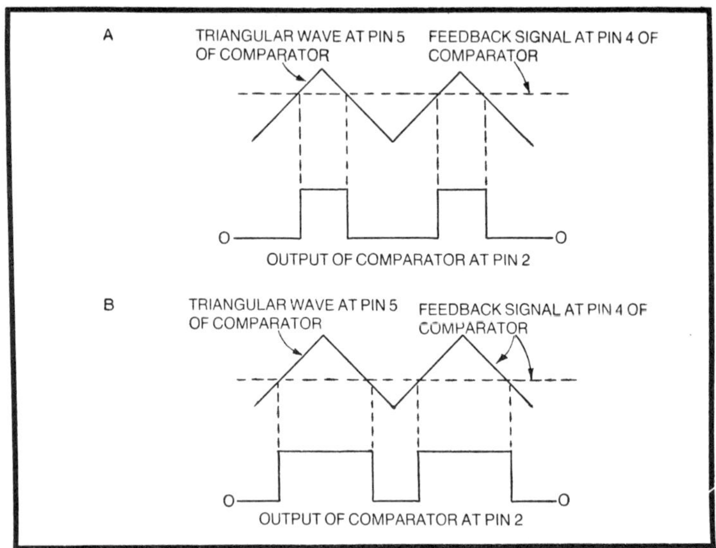

Fig. 13-18. Waveforms showing how the comparator responds to the sampled error signal. (A) Dc output level of regulator is high. This results in a low-duty-cycle switching wave, which tends to lower the DC output level of the regulator. (B) Dc output level of regulator is low. This results in a high-duty-cycle switching wave, which tends to raise the DC output level of the regulator.

action of the comparator is illustrated in Fig. 13-18. Although no voltage-reference source is directly associated with the comparator, the fixed-frequency, constant-amplitude, triangular wave may be said to perform this function. A 10V zener diode is provided in the output of the auxiliary power supply to ensure the amplitude constancy of the triangular wave.

The driving circuit for the 2N306 switching transistor encompasses the circuitry configured around the 2N6037, 2N6034, and MPS-U55 transistors. In Fig. 13-15, the driving circuit is actuated by the output of the comparator. The driving circuit is not merely a power booster or voltage level translator as in conventional regulators; this driving circuit generates a negative transient at the termination of *on* time, which speeds up removal of stored charge in the emitter—base region of the 2N6306 switching transistor. The negative transient is derived from the 10 μF capacitor. The drive current waveform is shown in Fig. 13-19.

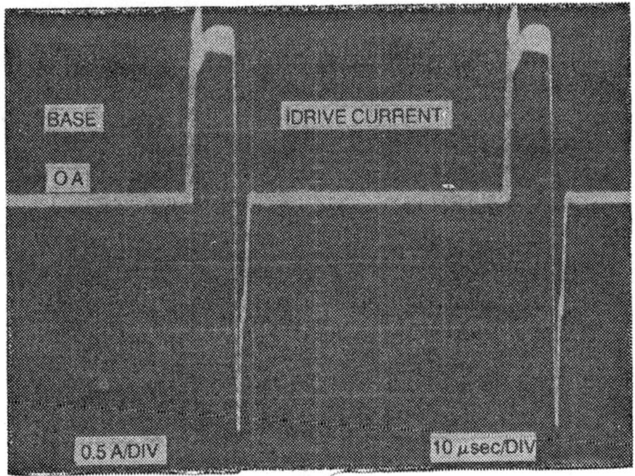

Fig. 13-19. Waveform of driving current for the switching transistor. Note the negative-going transient occurring at turnoff. This is purposeful, and speeds up reverse-recovery time of the switching transistor. (Courtesy Motorola Semiconductor Products, Inc.)

Actually, the 2N6037 and 2N6034 "transistors" are monolithic Darlington devices. Together, they form a push-pull arrangement for the production of the "charge-cleansing" negative transient. The primary function of the third device, the MPS-U55, is to limit the forward driving current to the switching transistor. Because of this provision, the base drive to the 2N6306 switching transistor is limited to about one ampere. (Excessive base drive to the switching transistor is a source of excessive dissipation in conventional regulators, and also slows fall time when the transistor is turned off.)

Current Limiting and Short-Circuit Protection

When an overload or short circuit occurs at the output, the normal feedback operation is overridden and a different mode of operation takes place to protect the regulator. While in this protective mode the regulator delivers 4A pulses with a duration of 30 μsec and a pulse repetition rate of about 1 kHz. Thus, the regulator is kept alive in an idling state. This protective technique features the resumption of normal operation once the short or overload is removed. Such overload protection is brought about in the following way:

Referring to Fig. 13-15, section 3 of the MC3302 quad-function IC module is used as a voltage comparator to sense the voltage drop developed across the 0.05Ω resistor connected in series with the 2N6306 switching transistor. When the output of the regulator is overloaded, this voltage becomes high enough to cause the comparator to change its conductive state, thereby discharging the 0.01 μF capacitor (C_X) in its ouput circuit. The capacitor requires time to recharge—in the meantime the comparator of section 4 is switched to its "high" logic state. This, in turn, is communicated through the 1N914 holdoff diode to the feedback comparator, section 2. This forms a closed loop, for comparator 2 now is forced to turn the switching transistor off. When this happens, the entire sequence of events repeats, and as long as the overload exists at the output of the regultor, this low duty-cycle pulsing persists. The feedback comparator of section 2 is, however, ready to resume normal operation once the overload is removed. During normal operation of the regulator the feedback comparator is undisturbed by the current-limiting circuitry because of the isolating action of the 1N914 holdoff diode.

14

Applications of DC Switching-Type Power Supplies

It is interesting to note the way that producers of consumers products have delved into the basics of switching-type supplies to develop circuits with somewhat different uses of components than their counterparts in instrumentation, transportation, space applications, etc. The unique aspect of the consumer's market, is, of course, cost.

AN SCR SWITCHING-TYPE
REGULATOR FOR COLOR TV RECEIVER

The color TV supply shown in Fig. 14-1 makes use of the fact that SCRs can simultaneously provide both rectification and control. Instead of a conventional unijunction oscillator, this design uses the so-called *programmable* unijunction transistor, in this instance a MPU131. (This device performs much the same function as the ordinary unijunction transistor in Fig. 3-3). The MPU131 is connected as a relaxation oscillator which actually never runs freely, but is reset every time the AC line voltage crosses zero; in other words, the oscillator is snychronized to the line frequency. No special circuitry is needed to accomplish this, for it is brought about by the 120 Hz output ripple from the lightly filtered full-wave rectifier formed by the 1N4003 diodes.

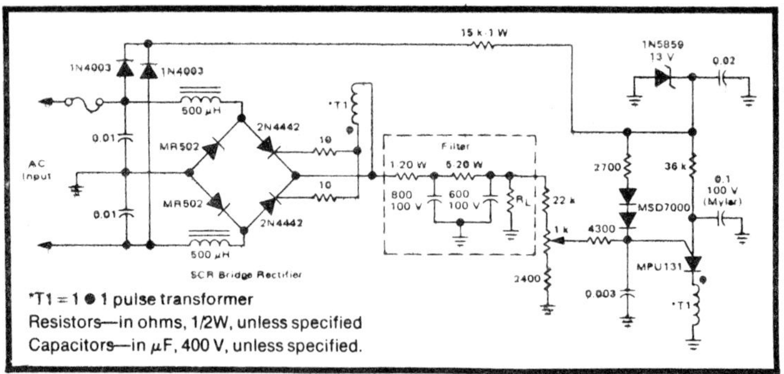

Fig. 14-1. An SCR switching-type supply for color TV receivers. This supply provides a nominal 80V output at currents up to 1.5A. About 200 mA is the minimum load, because of SCR holding current. (Courtesy Motorola Semiconductor Products, Inc.)

Whether the programmable unijunction transistor fires early or late in the AC cycle depends upon the DC voltage at its gate, which is obtained from the output of the supply. (Note that the TV set is represented by R_L). If the DC output voltage starts to increase, the MPU131 starts to fire later, delaying the firing of the SCRs to lower the average DC output level. If the output voltage starts to decrease, the SCRs fire sooner. A given SCR can be triggered into conduction only during that portion of the AC cycle when its anode is positive relative to its cathode. Thus, the full-wave-rectifying property of the diode/SCR bridge is retained.

The usual output LC filter in a switching-type regulator is absent here. The filtering is provided instead by an RC network, despite the obvious I^2R power loss. This technique is cheaper to implement than the usual inductor—capacitor combination; however, small LC line filters are provided in order to minimize noise on the power line due to SCR turnon pulses and other noise generated by the TV set itself, such as the vertical and horizontal frequencies. This supply can provide 80V at 2% regulation over a line-voltage range of 105—140V. Current capability is 1.5A.

TEMPERATURE REGULATOR
USING ZERO-VOLT-SWITCHING TECHNIQUE

The circuit shown in Fig. 14-2 is a switching-type power supply in which the load is an electric heater. The feedback

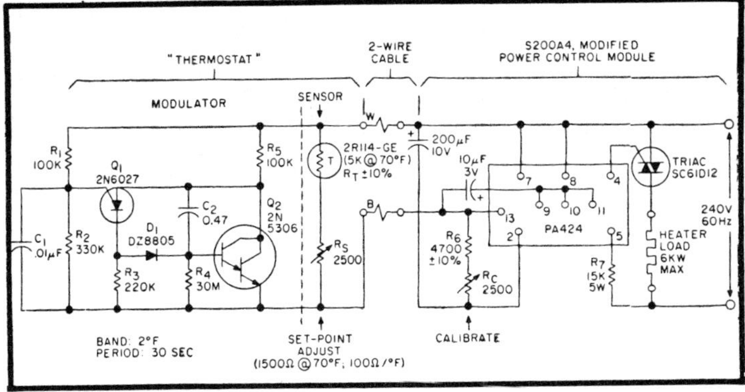

Fig. 14-2. Temperature regulator using zero-voltage-switching technique. (Courtesy General Electric, Semiconductor Products)

path consists of air space, and the sensor is a thermistor. The novel feature of this arrangement is that the electrical power to the heater always comprises an *integral number of cycles*. When more power is needed, the pulse trains of voltage are applied to the heater and made to contain an additional cycle. The beauty of this switching technique is that virtually no RFI and EMI is generated. Accordingly, you will not see the customary high-frequency-filter networks so prominent in conventional thyristor control and regulating circuits—those using phase control of *individual* cycles. The heart of the scheme is the General Electric PA424 module, which generates trigger pulses for triacs only when the line voltage crosses zero, and then only when commanded to do so. Therefore, the triac always delivers an integral number of cycles to the load. The operation of the temperature controller is easier to grasp if one thinks of the PA424 as simply another gate-trigger device, to be used in place of a *diac* (a 4-layer diode), a unijunction transistor, or a pulse transformer. In the functional diagram of Fig. 14-2B, the zero-crossing detector can deliver a forward-bias pulse to the base of the output Darlington stage. In turn, a gate firing pulse is available from terminal 4. Operation is governed by the signal appearing at terminal 13. If transistor Q_1 is held in its conductive state, no pulses appear at the output. The power supply consists of diode D_6 in conjunction with an external capacitor, such as the 200 μF unit shown above in Fig. 14-2. Whenever transistor Q_1 is

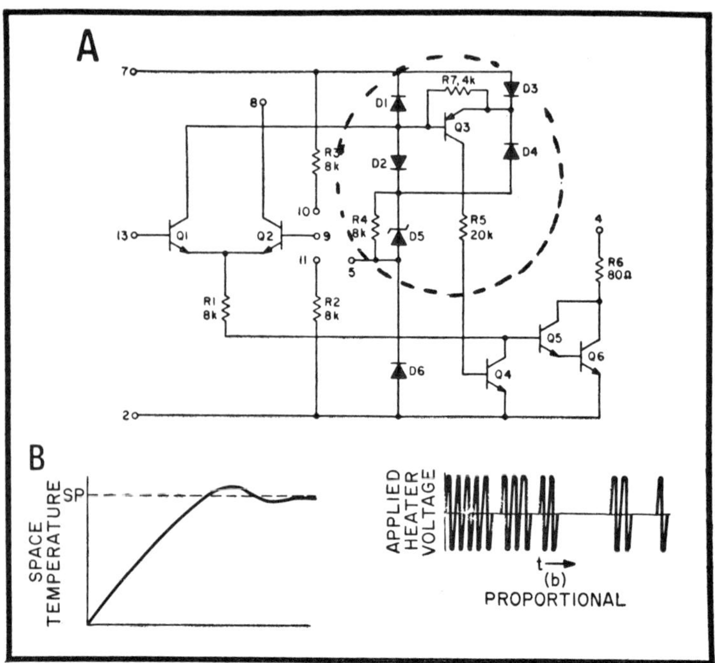

Fig. 14-3. Zero-voltage switching is accomplished using the PA424 zero-crossing switch. The output of the circuit shown in Fig. 14-2 consists of a series of complete cycles, and so produces minimal switching noise. (Courtesy General Electric Semiconductor Products)

taken out of conduction, a gate-firing pulse appears at terminal 4 at the very next zero-crossing of the line voltage.

The modulator consists of a programmable unijunction transistor (a 2N6027) in a relaxation oscillator having a period of about 30 seconds. This slow waveform is superimposed upon the signal sensed by the thermistor at terminal 13 of the PA424. This simulates a temperature excusion of about 2°F and causes the system to *anticipate* the temperature established adjustment of R_S. This results in less overshoot and undershoot (Fig. 14-3) of the equilibrium temperature than would be the case without such modulation. Thus, this electronic thermostat displays much less thermal lag than a typical bimetallic switch

Integral-cycle switching is one of the newer regulation concepts. Thus far, its implementation has been largely limited to 60 Hz applications. This scheme has interesting potential for high-frequency supplies, where it may emerge as a possible means of reducing electrical interference.

MOTOR SPEED REGULATION
BY PHASE-CONTROLLED SWITCHING

The simple circuit shown in Fig. 14-4 provides adjustment and regulation for the speed of DC and universal-type motors. Only a few parts are involved, but there is feedback, sensing, comparison, and pulse-time modulation occuring in this circuit, much like the servo action in complex regulators.

The armature of the motor spins in a magnetic field, and such motor action is accompanied by an internal *generator* action. The armature develops a counter-emf that opposes the flow of current into the motor. Under no-load conditions, the counter-emf approaches the value of the actual voltage applied across the armature. This means that only a relatively small internal voltage drop provides the force to push current through the motor. The current is not zero, since the motor needs some current to overcome its own frictional losses. Under heavy loads, the counter-emf decreases and larger currents are passed to develop electromagnetic torque sufficient to carry the load. Since the armature is connected in the gate/cathode circuit, the faster rotation which occurs

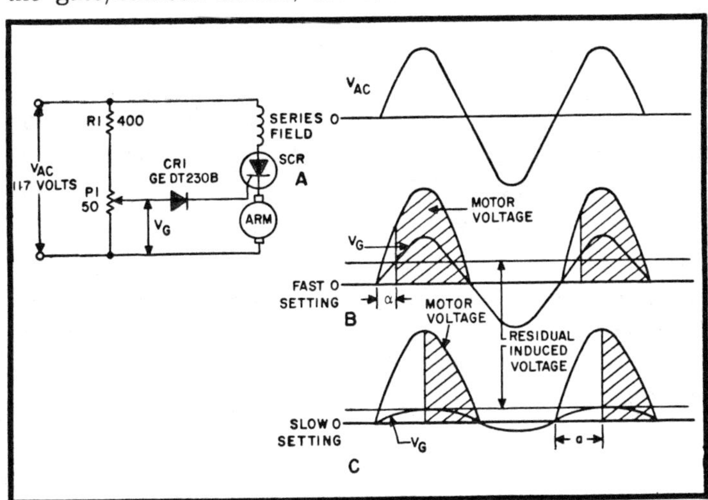

Fig. 14-4. Application of phase-controlled SCR to regulation of motor speed. (Courtesy General Electric, Semiconductor Products)

under light loads inhibits firing of the gate because of the bucking effect produced by the counter-emf on the cathode. The motor speed is goverened by the gate voltage obtained from the R_1/P_1 network.

Delayed firing corresponds to less average current through the motor, so its torque is decreased. This technique is used to reduce the motor speed if it tries to increase due to a relaxation in loading. Under greater shaft loading, the speed of the motor tends to decrease, lowering the counter-emf and enabling the SCR to fire earlier. The higher average current which results counteracts the drop in speed.

The SCR and the motor must be compatible in current ratings, which is to say that the SCR must have *sufficient* current capability. Although this simple regulation technique has been applied widely for the control of small motors, it can be applied to the control of larger motors. It does not, however, exhibit stable low-speed characteristics, and additionally a large installation would also require adequate over-current protection.

MOTOR-SPEED REGULATOR
WITH FULL-WAVE CONTROL

Half-wave control of motor speed has enjoyed wide popularity. The circuits in Figs. 14-4 and 14-6 feed the motor via half-wave rectification. There is no transformer secondary in these circuits, so core saturation by the DC component of half-wave rectification is not a problem. The motor inductance, together with the free-wheeling diode, also tend to make the torque less pulsating then it might otherwise be. On the other hand, a full-wave supply of power to the motor results in much smoother operation at low speeds, and generally extends the control range. A circuit for accomplishing this is shown in Fig. 14-5. The speed-regulating principle remains the same as for the half-wave circuits.

A bridge rectifier is needed to route the current so that it is always passed through the motor in the same direction. This method is particularly advantageous when full-wave performance is desired, but the power requirements of the motor rule out the use of a triac. No transformer is shown, but one could be used for the sake of isolation.

244

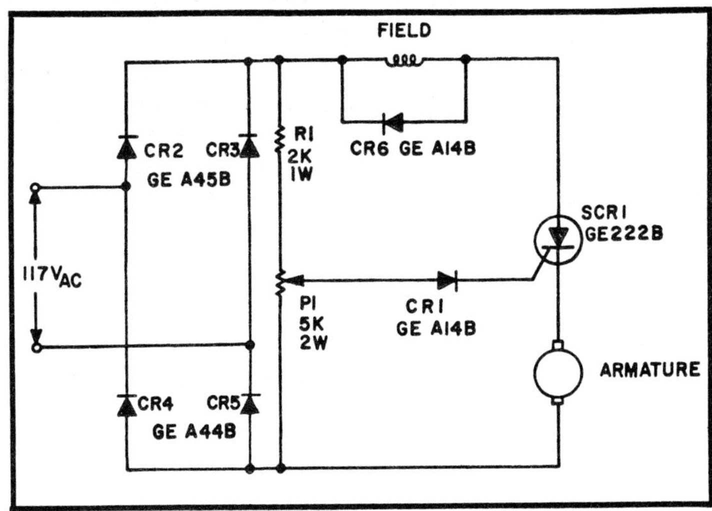

Fig. 14-5. A full-wave speed regulator. (Courtesy General Electric Semiconductor Products)

Variations of this full-wave scheme are also possible. One, in particular, simply places the speed-regulating unit in series with the motor and the AC power line. This means that the primary function of the regulator is to maintain a constant average value of AC voltage across the motor. While this technique does exert some control over the speed of the motor, it does not respond well to variations in motor loading. However, by modifying the control circuit, it would be possible to incorporate a tachometer and optoisolator to feed back speed information.

USING A SWITCHING-TYPE POWER SUPPLY FOR SPEED CONTROL

Although the use of solid-state electronics for motor control is a well-known technique, it often happens that only those directly involved are aware of the many things that can be accomplished. By means of sensing, feedback, and pulse manipulation—the very techniques used in switching-type power supplies—the starting, speed, and torque characteristics of motors can be altered at will. An example is the synchronously controlled motor circuit shown in Fig. 14-6, in which the DC motor is caused to maintain its speed with the

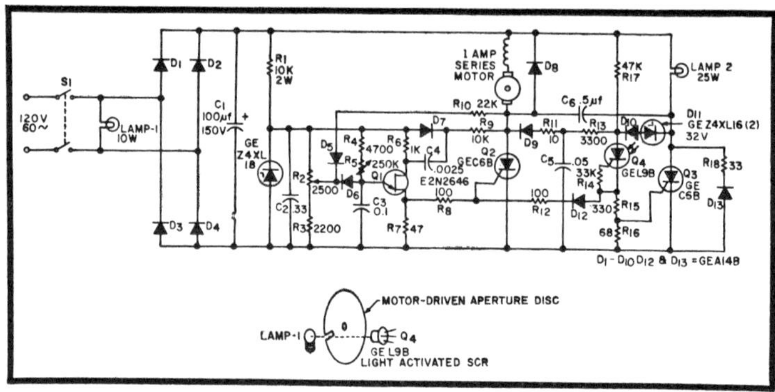

Fig. 14-6. Precise regulation of DC motor speed with a switching-type power supply. This control technique gives the series-type DC motor the same speed characteristics as an AC synchronous motor. (Courtesy General Electric Semiconductor Products)

same constancy as an AC synchronous motor. At the same time, the starting and pull-in torque remain high.

In this circuit, the main SCR is Q_2, which receives turnon pulses from the unijunction-transistor oscillator configured about Q_1. Commutation or turnoff action is imparted to this SCR by Q_3 which in turn is controlled by the light-activated SCR (Q_4). Light coming through the aperture of a motor-driven disk tends to turn Q_2 off, and the oscillator tends to turn it on. If the motor tends to run slower than its synchronous speed—the rate of turnon pulses delivered to Q_2—the duration of the pulses will be lengthened, so that the average current passing through the armature and field is increased. This increases the electromagnetic torque and the motor accelerates. If it should accelerate to a speed in excess of its synchronous speed, the current pulses from the main SCR become too short to maintain such speed, and the motor falls back to its regulated rotational rate. If the load on the motor varies, the width of the pulses delivered by the main SCR vary in order to provide just enough average torque to the motor so that its pulse turnoff rate equals the pulse turnon rate.

In order that the motor action will be critically damped, delaying circuitry is inserted in the feedback path. Components R_2, R_3, R_{10}, D_5 and D_6 represent one such damping network. Diode D_8 is equivalent to the free-wheeling diode in ordinary switching supplies, and it provides a current path for

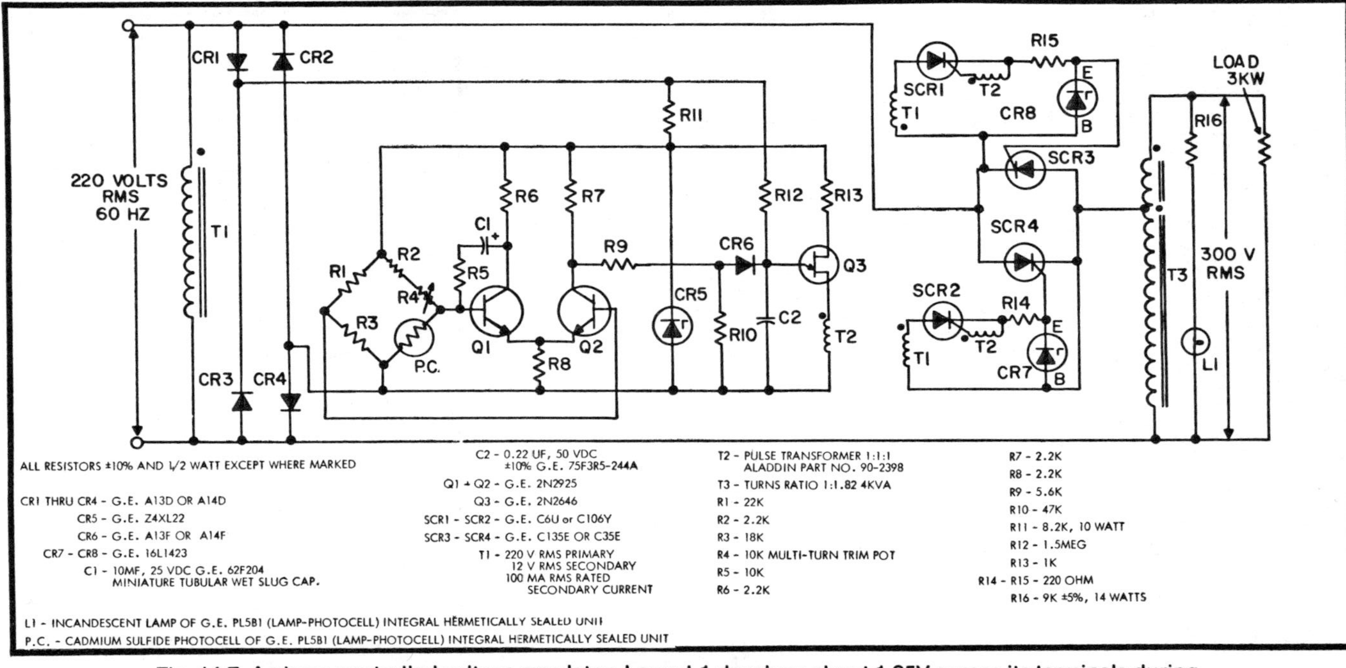

Fig. 14-7. A phase-controlled voltage regulator. Lamp L1 develops about 1.95V across its terminals during steady-state operation. (Courtesy General Electric Semiconductor Products)

sorted energy in the motor inductance during SCR *on* time. Note that in this circuit, *frequencies* rather than voltage levels are compared; therefore, no zener or other voltage-reference source is needed. This scheme has application to recorders and other instrumentation, particularly when operation must be obtained from battery power. (The input bridge rectifier can operate from DC as well as AC).

PHASE-CONTROLLED AC VOLTAGE REGULATOR

A tried and proven systems is shown in Fig. 14-7. It is a flexible arrangement and readily lends itself to the use of other solid-state elements. For example, General Electric markets an extensive line of SCRs suitable for kilowatt operating powers. The basic circuit can also be scaled down to lower powr levels. Unlike most of the regulators in this chapter, this one regulates the RMS level of its AC output, rather than providing voltage-stabilized DC. Its applications have included constant heat for controlling chemical and industrial processes, photocopying machines, AC motors, and special military uses. It is readily converted to an RMS AC current regulators. (See Fig. 14-8). A lighter-weight 400 Hz version is possible with an appropriate transformer and a few other modifications.

The power-handling portion of the circuit consists of output autotransformer T_3 and power thryristors SCR_3 and SCR_4. The

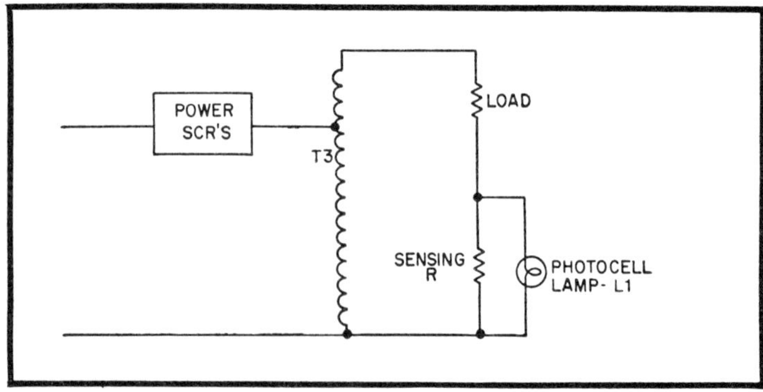

Fig. 14-8. A slight change in the incandescent-lamp circuit produces a current regulator. In this variation, the RMS current to the load is regulated. Again, lamp L1 develops about 1.95V across it during steady-state operation. (Courtesy General Electric Semiconductor Products)

gates of these devices are controlled by the smaller "pilot" SCRs. An optical-feedback path is provided by the lamp/photocell unit, designated respectively as L_1 and P.C. The cadmium photocell is connected in the arm of a resistance bridge. The resistance of the cell governs the base–emitter bias of transistor Q_1, which is the input stage of the differential amplifier. The conductive state of Q_2 determines how quickly capacitor C_2 can attain sufficient voltage to fire unijunction transistor Q_3, which provides the triggering pulses for the pilot SCRs. The feedback loop is thus able to regulate the output voltage by phase control of each half-cycle. Although Q_3 is connected as a relaxation oscillatior, it is *not* free-running—rather, it is synchronized to the power line via the ripple in the unfiltered DC impressed across its bases. Every 1/120 second this oscillator is *reset* and C_2 commences a fresh charging ramp. Interestingly, even though the output waveform in nonsinusoidal, it is the RMS value of the output voltage which is regulated, because the luminous intensity of the lamp is proportinal to the RMS current through it, and therefore to the RMS voltage applied to it.

Index

250